THEORISING

A Primer for Soci

Edited by
Johanna Ohlsson and Stephen Przybylinski

ı

BRISTOL
UNIVERSITY
PRESS

First published in Great Britain in 2023 by

Bristol University Press
University of Bristol
1–9 Old Park Hill
Bristol
BS2 8BB
UK
t: +44 (0)117 374 6645
e: bup-info@bristol.ac.uk

Details of international sales and distribution partners are available at bristoluniversitypress.co.uk

Editorial selection and editorial matter © 2023 OHLSSON, © 2023 PRZYBYLINSKI, individual chapters © their respective authors 2023

British Library Cataloguing in Publication Data
A catalogue record for this book is available from the British Library

ISBN 978-1-5292-3222-6 paperback
ISBN 978-1-5292-3224-0 ePub
ISBN 978-1-5292-3223-3 ePdf

Cover design: blu inc
Front cover image: stocksy/NadineGreeff
Bristol University Press uses environmentally responsible print partners.
Printed and bound in Great Britain by CPI Group (UK) Ltd, Croydon, CR0 4YY

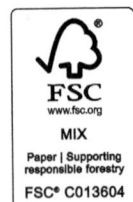

FSC
www.fsc.org
MIX
Paper | Supporting
responsible forestry
FSC® C013604

Contents

List of Figures and Tables

Figures

Tables

Notes on Contributors

Darren McCauley is Professor of Environmental and Social Justice (Law) at Newcastle University. He is also research group convenor for Environmental Challenges and Law. His mixed methods research agenda focuses on investigating what a just transition to a low-carbon future looks like from a global perspective. This has involved funded empirical research in the Arctic, sub-Saharan Africa, Southeast Asia, Europe and the United States. He works closely with international organisations to co-develop research and ensure maximum impact from his work.

Don Mitchell is Professor at the Department of Human Geography, Uppsala University. His interests in geography are broad, but are tied together by a commitment to historical-materialist analyses (with an emphasis on the historical part of the equation), and a strong – though not exclusive – focus on understanding the political-economic determinants of landscapes, cultures and urban public spaces.

Johanna Ohlsson is Assistant Professor at the Department of Human Rights and Democracy, University College Stockholm, and Researcher in ethics at the Institute for Russian and Eurasian Studies, Faculty of Social Sciences at Uppsala University. Her research interests cover human rights, sustainability, models of justice and political processes.

Stephen Przybylinski is Assistant Professor at the Department of Geography, Environment and Spatial Sciences at Michigan State University. He is an urban and political geographer broadly interested in the ways in which liberal democracies both enable injustices and how such political systems mediate responses to injustices within their frameworks.

Roman Sidortsov is Associate Professor in Energy Policy at Michigan Technological University, and Senior Research Fellow in Energy Justice (SPRU – Science Policy Research Unit), University of Sussex Business School. Dr Sidortsov has a diverse international background as an educator, researcher, consultant and practising attorney. He has developed and taught law and policy courses ranging from Renewable Energy and Alternative Fuels to Administrative Law at Vermont Law School, Irkutsk State Academy of Law and Economics in Russia and Marlboro College Graduate School's MBA in Managing for Sustainability programme.

Tracey Skillington is Director of the BA (Sociology) and Chair of the Undergraduate Committee in the Department of Sociology and Criminology at University College Cork. She is currently researching issues of justice that arise in relation to global climate change, cosmopolitanism and transnational democracy. Dr Skillington's research interests include critical theory, climate justice, human rights, global justice, intergenerational inequalities, border practices, sociology of the body, European identity, cosmopolitanism, collective memory, trauma and models of democracy.

Corine Wood-Donnelly is Associate Professor of International Relations and the High North at the Faculty of Social Sciences at Nord University. She is also a researcher at Uppsala University where she is the Scientific Coordinator for the EU-funded project JUSTNORTH (GA 869327). Dr Wood-Donnelly is an interdisciplinary researcher in International Relations and political geography and specialises in governance and policy of the Arctic region.

Preface

Theorising Justice began as a collaborative effort at understanding the concept of justice to problems central to the social and environmental sciences. This volume comes out of JustNorth, a project which received funding from the European Union's Horizon 2020 Research and Innovation Programme. JustNorth is dedicated to examining the potential for just, ethical and sustainable development in the Arctic.

The contributing authors come from the disciplines of human geography, sociology, ethics, legal studies and political science. The grant, and now this volume, illustrate how truly multidisciplinary this work has been. We believe this diversity has greatly improved the development of these chapters. Early versions of the chapters began as discussion papers, which were presented in our research seminars and have served as starting points for our work in developing the overviews, assessments and analyses throughout these chapters.

We are grateful to have received funding from the European Union's Horizon 2020 Research and Innovation Programme under Grant Agreement No 869327.

Johanna Ohlsson and Stephen Przybylinski
April 2023

Introduction

Stephen Przybylinski and Johanna Ohlsson

Introduction

This primer on justice intends to illustrate the pluralism present within and between justice theories to illustrate their application to the social sciences. Promoting the pluralism of justice theories in themselves is not our objective, though it is a secondary benefit. Rather, the volume encourages readers to analyse the potential insights that differing theories of justice may offer them for explaining conditions of injustice. Taken as a whole, *Theorising Justice* promotes the value of seeing variances within justice approaches, by considering not only the philosophical and normative underpinnings of justice theories, but also how justice is conceptualised in respect to more traditional topics of focus for social and political science researchers concerned with social and environmental injustices.

A central pursuit for many social and political science scholars has been the realisation of social or environmental justice. While these general concepts are not unfamiliar to the social and political sciences or public policy scholarship, reconceptualising how justice may be differently theorised to explain social or environmental concerns has been less central to this scholarship. This is not to say that justice theorising has been absent from social and political science scholarship. The term justice will remain a central framework to analyse social and environmental problems. Yet, justice theorising often and understandably derives from its root in liberal theories, reinforcing and thus potentially reducing the gains that new and competing analytical approaches to justice theorising outside the liberal tradition may achieve for traditional social and environmental sciences. Which begs the question, if justice theories are potentially limited by their liberal foundations, what do alternate theories of justice help us see about the social and environmental conditions of the world? And, equally important for the purposes of this volume, how do justice theories rooted within and outside the liberal tradition help social and political science scholars offer new explanations for the conditions they study?

Justice theories provide a set of explanations necessary for evaluating conditions of injustice. Justice theories help explain actually-existing circumstances of social, political or environmental injustices. As concepts, they help lay out ideal circumstances which assist in evaluating, for instance, whether certain policies addressing given social or ecological conditions are fair, equitable or democratic. But justice theories also bolster social and political scientists' positions on what policies or relations *ought* to look like in a just society. For, conceptualisations of justice rely upon moral reasoning to assess the very social, political and environmental processes leading to unjust conditions. The moral reasoning inherent to more normative and philosophical scholarship, whose purview traditionally lies outside of the social sciences, works to bridge the concept with those of the social sciences. And it is exactly for this reason that conceptualisations of justice become more powerful when assessed and applied within social and political science research. Justice theories may be just as useful when they analyse how social, economic or ecological injustices are played out in actually-existing conditions in the world, as they are when they seek to explain how such conditions ought to be changed, by identifying the ideal circumstances needed to realise more just relations.

At the same time, normative arguments theorising what should be done to bring about more just relations can be limited by the conditions surrounding any given problem. What may be theoretically understood as just within one set of social, political or ecological circumstances may not be just in another. For example, equally limiting carbon emissions of all countries across the world may not be just when understood through the historical context of how certain societies have premised their development on carbon exploitation. Identifying what is just in theory becomes more complicated when its principles are used to evaluate the actual material and social conditions of a given society or context. That the process of analysing and assessing what constitutes just or unjust relations becomes more difficult when examining a given case does not mean that justice theorising is not useful, however. Far from it. There is much analytical purpose in identifying alternative ways of assessing injustice in social and ecological problems. For, there are multiple ways of thinking about what justice or 'just' relations can mean for different groups and non-human species. From our perspective, one is best equipped to conceive of what is just or unjust when one has a clearer view of how justice theories meet the material realities of the world.

This is why social and political science research makes such an important contribution to justice theorising. By supplementing empirical descriptions of injustice with normative assessments of the conditions maintaining them, researchers may develop more nuanced and incisive analyses to affect the changes needed to realise more just conditions or circumstances. We argue that there is much value in assessing the potential application and utility of many theoretical approaches to justice. For, no theory of justice is equal to another; each come from different epistemic positions and political orientations, and each can evaluate quite a variety of social and ecological circumstances. To arrive at which set of values and arguments are better suited for evaluating a given injustice, we argue that this is best done by analysing the potential strengths and weaknesses of a variety of justice theories. Collectively, the overviews presented throughout the following chapters illustrate well the epistemic, ontological and political diversity of justice theorising. Whether new to theories of justice or well-versed in its many forms, the aggregate of approaches surveyed herein offers students and researchers the opportunity to (re)locate the value of justice theorising.

Framing justice theories and approaches

This volume represents one attempt at creating connections between the theoretical foundations and the actually-existing conditions of justice. Importantly, the book provides overviews of philosophical and normative justice theorising alongside social and political science analyses of injustices. We find this arrangement useful, not only to illustrate how normative justice theorising helps frame social-environmental analyses, but also to highlight what shortcomings exist when using such theories to explain everyday problems. For, approaches to justice theorising are incredibly diverse. Some theories and approaches understand justice as an ideal, while others see it as a process or an achievable or measurable goal. Justice is also a value, a set of principles, a practice, a relationship, a condition, a concept, a state of being, or a state of becoming. Some of these understandings of justice might be mutually reinforcing, while others are mutually exclusive. How one researcher approaches justice can depend on factors such as which disciplinary domain they are trained in, what ontological and epistemological positions they see a set of issues through, in addition to their political or ideological orientation more broadly. Moreover, while justice indicates a universal (or perhaps better, generalisable) set of conditions or practises, the specific content and meanings of justice are always forged in particular historical, geographical, social, cultural and political contexts, and the relations of power that shape these. We see the diversity of approaches to justice theorising as both a challenge and an opportunity, and see this volume as a guide to illustrating the opportunities available within the most prominent theories of justice.

Our intent with this volume is not to settle for one or even a few definitions or conceptualisations of justice. Nor is its purpose to articulate some meta-theoretical position on justice. Instead, we find it more useful to identify what the strengths and limitations, and thus applications, are of the many different approaches to justice selected for the volume.

The chapters selected for the volume were chosen to create a collection illustrating the robustness of justice traditions. Given the extent of justice theories, however, we have chosen chapters surveying the most well-defined and, in some cases, quickly emerging traditions represented in social and political science scholarship on justice today. Mainstays of social and political science scholarship on environmental and climate justice, for instance, are paired with the rapidly expanding approaches to energy injustices. But despite the range of topics and approaches among the chapters, we find a throughline emerging from, or critically disconnecting with, liberal theories of justice. As we discuss at length in the conclusion, liberal justice theories dominate the scene, by functioning either as the root of justice principles *or* as a springboard for critiques and alternative approaches to the dominance of the liberal tradition. With this in mind, each chapter details the conceptual influences related to each tradition, examining debates within, and critiques of, each approach to justice. Through these overviews, we hope to illustrate to readers that, while the liberal tradition continues to shape much justice theorising, there is no one approach or tradition that best analyses justice.

Despite the general structure guiding these chapters, authorial voice is not lost in chapter overviews and readers will likely sense differences in author positions. We find this does not detract from, but enhances the volume's overall contribution to justice theorising in the social sciences. To create a more cohesive narrative throughout the chapters, then, we have identified thematic categories to help readers think about, and make connections through, the diversity of perspectives herein. We call these categories the 'forms', 'aspects' and 'realms' of justice. Necessarily simplified, we adopt these categories for pedagogical purposes, to help readers critically approach a given justice tradition from multiple perspectives. The first of these categories, the 'forms' of justice, represent five core features helping to distinguish justice from other moral and political ideas. We identify these key forms as: *substantive justice, procedural justice, distributive justice, retributive justice* and *recognitional justice*. While many other 'forms' of justice exist, such as restorative, rectificatory and reparative justice, for instance, we have focused on these five because they are the most prominent across justice literatures and have sparked the most debates, not only in legal theory but also within philosophical discourse and applied justice theory approaches.

We summarise the five forms of justice as follows.

Substance, or *substantive justice*, pays attention to the core of justice and has historically been understood and approached in several ways. Legal approaches to substantive justice understand it either as a sufficient degree of fairness or as the substance of the law, for instance, whether the rules in themselves are just or not. Other more normative approaches understand it as the concern with, and critique of, the very ideals of justice.

Procedure, or *procedural justice*, as a legal concept is commonly understood as the form of the law or practice which should, and must, be applied equally to all. It is also commonly understood as connected to political decision-making on various administrative levels or judicial processes, such as trials or appeals in courts, and so on. Justice in this sense comes through the procedures used to determine how benefits and burdens of various kinds are allocated to people. Yet, this allows for different outcomes, and procedural justice can also be understood, as John Rawls (1999 [1971]) suggested, as either *perfect* (always producing a just outcome), *imperfect* (likely to produce a just outcome) or *pure* (no way to determine if the procedure will lead to a just outcome).

Distribution, or *distributive justice*, is centred on either the justness of allocation or the outcome of procedures. This includes various approaches guiding or governing the dispersal of goods, benefits or burdens among members of society. How societies are organised (for example, their economic, political and social frameworks in terms of law, institutions and policies) results in different approaches to distribution, which fundamentally affects people's lives. Such frameworks are not static, and the discussion on how they *are* and *should* be set up – and both legally and morally justified – is one of the core questions of distributive justice.

Retribution, or *retributive or corrective justice* is primarily concerned with the punishment of wrongful acts. This might sound straightforward, but several features of retribution, such as the notions of desert and proportionality, the normative status of suffering, and the ultimate justification for retribution have long been debated. Indeed, questions of retribution are closely tied to questions of substance and procedure.

Recognition, or *recognitional justice*, pays attention to how differences among peoples and groups affect their treatment relative to others. Recognitional justice emphasises why cultural differences matter for fair treatment, especially for how the recognition of such cultural differences may enhance social equality more broadly. This set of approaches is thus critical of redistributive theories of justice which do not attend to cultural forms of recognition, forms that overlook cultural disrespect or humiliation as being less important or even unrelated to economic disadvantage, for example. From this perspective, a just society is a society in which conditions exist where individual dignity may exist for all people.

While these five forms of justice offer one useful way of categorising justice discourse and scholarship, we also find it useful to consider other general ways of framing what justice can be about. To supplement the forms, we identify five 'aspects' of justice. These aspects focus on the 'who, what, where, when and how' of justice, as developed by Allison Jaggar (2009), and should not necessarily be seen as mutually exclusive to the five forms.

We identify the five aspects of justice as follows.

The *subject* (or 'who') of justice points to whom justice principles or concerns are applicable – that is 'who counts': Individuals *and* groups? Men *and* women? Citizens *and* strangers? Humans, animals *and* nature? Children? Future or past generations? Over the history of philosophical, as well as practical, development, the subject of justice has been a source of considerable debate and social struggle: until quite recently, for example, women were not subjects of justice except insofar as they were subjected to the dominance of fathers, husbands, brothers or other men. The status of animals and nature as subjects of justice remains controversial. Some traditions place the group ontologically prior to the individual while others do the opposite.

The *object* (or 'what') of justice concerns the entities to which justice claims are addressed: institutions, nation states, social processes, and so on. In John Rawls' (1999 [1971]) influential theory of justice, the object of justice is the 'basic structure' (the key institutions of society), but not, in his original formulations, the family; by contrast, the prominent feminist theorist Susan Moller Okin (1989) argued forcefully that the family had to be the object of justice, a key constituent of the 'basic structure' (see Chapter 4). For Marxists and anarchists, the object of justice is, in the first instance, the mode and relations of production (see Chapter 5). In the Capabilities Approach, it is the condition of possibility – the capability – for (to take only one example) healthy life (see Chapter 8). More specifically, in liberalism, the object of justice is primarily centred on *(re)distribution*, while feminism and other approaches have sought to expand it to also include *recognition* and *representation* (Fraser, 1997).

The *domain* (or 'where') of justice concerns the arena (or geographical or political scale) within which justice claims are made and addressed. Is it the nation state (the assumed domain of much, though not all, liberal justice theory) or is it more global than that (as posited by many cosmopolitan and feminist theorists, as well as in the discourse of 'global justice')? Is the domain of justice the community, village or region? In other words, how are universality and particularity juggled – and at what scale – in theories of justice? Much contemporary work, as will be seen throughout the following chapters, adopts a complex and nuanced understanding of the intersection of different geographical scales, the role of the nation state and the relationship between the universal and the particular.

The *social circumstances* (or 'when') of justice concern the conditions under which principles of justice become salient. For a long time, 'most modern [liberal] philosophers agreed that the circumstances in which the principles of justice were salient were those of moderate scarcity' (Jaggar, 2009, p 3), but radical thinkers might very well disagree, arguing, for example, that it is in production rather than distribution that questions of justice arise even when societies of abundance are imagined. Feminists argue that the principles of justice arise in relation to the operation of patriarchy, whatever the relations of production and distribution in society. Postcolonial theories of justice look to the operation of geopolitical (among other forms of) power to understand when principles of justice come into play.

Finally, the *principles* (or 'how') of justice concerns *procedure* as discussed earlier, but also encompasses substantive principles in that any procedure must be assessed in terms of outcome and/or process A core distributive principle, for instance, is often phrased as 'a just distribution justly arrived at' (Harvey, 2009 [1973], p 98). For cosmopolitan theorists, this aspect of justice concerns the principles by which 'the right to have rights' are determined within a global society marked by extensive migration across borders (see Chapter 3).

Finally, given that the traditions discussed in Part I and Part II are placed on different analytical levels and have different foci, we have inductively developed a third conceptual layer for comparing between and across the chapters in Part II. We call this third category 'realms'. Each of the applied fields presented in Part II can be understood in relation to a set of realms defined by:

- the *temporalities* they work in;
- the *scope or scale* of the phenomena they examine;
- their specific *locus of concern*;
- and the *source of the harms* they are concerned with.

We find the realms an additional means of situating or identifying features of justice beyond the forms and aspects that researchers may find useful to begin their inquiries. We stress, however, that by introducing these categories, we are not exhausting the ways that justice theories can be understood. Instead, we intend the categories to promote new ways of making connections in, and finding relevance with, justice theories, particularly in the application to the social and political sciences.

Chapter contents

Each chapter in the volume engages with the forms, aspects and realms of justice differently. For instance, some chapters focus more on the substantive and procedural forms of justice, such as in the liberal justice chapter. Others heavily centre on the aspects of subjectivity, or the 'who' of justice, for instance, such as found in the feminist, postcolonial and Indigenous justice chapters. It should be noted that the chapters are not forced into these categories. Rather, the chapters emphasise the conceptual development and debates of these traditions, to which these categories may be used to help readers systematise what they find valuable in a given tradition. With that in mind, we have organised the volume into two parts and provide summaries to each chapter here.

Part I: Politico-philosophical and normative traditions of justice

It is a commonplace to begin a volume on justice with an overview of the liberal tradition. But so too is it reasonable. While liberal theories of justice have developed over centuries, the more recent approaches detailed throughout this volume are derived from, or are critically engaged with, concepts germane to the liberal tradition, or what we refer to as the 'mainstream' of justice theorising. It is difficult to provide an exhaustive overview of liberal theories of justice within the space of one chapter. As such, Chapter 1 attends to how a few key liberal theorists have pursued justice in its substance and how rights discourse subsequent to these key theories assesses the function of rights procedurally. The influential work of Thomas Hobbes, Immanuel Kant and John Rawls, for instance, all identify what constitutes the substance of justice while at the same time prescribing how justice may be pursued procedurally, such as through redistributive schema which balances economic and political equality among members of a society. Key to many theories and traditions, and liberalism more broadly, is a concentration on the value and function of legal and moral rights, often in an idealised fashion. Yet, as Chapter 1 also shows, more recently scholars have questioned who, exactly, the subject of liberal justice has been, continues to be, and ought to be. That is, who has been and still is the rights-bearing subject of justice within liberal societies? Given that White, property-owning males exclusively constituted the liberal subject for centuries, scholars have argued that we should reconsider not only who can be a subject of liberal justice, but also how those excluded from liberal justice can become legitimate subjects of justice in the first instance and whether liberal ideals are the means through which justice can be achieved for those excluded from these premises. The chapter's attention to the forms of both substance and procedure coalesces with a more recent focus on the subject and object of justice within liberal theories and their critiques.

Chapter 2 details the dominant libertarian theory of justice, one rooted in individual liberty and natural rights. In this sense, the 'mainstream' libertarian approach overlaps with liberal theories of justice, by prioritising the protection of property rights and self-ownership, resulting in a narrow conception of justice that disregards equality. Where mainstream libertarian theories differ, however, is that they focus on the protection of individual rights and the enforcement of contracts with no regard to equality in outcomes. The second part of the chapter shifts to examine left-leaning critiques of mainstream libertarianism and offers alternative perspectives on just relations, including a re-examination of historical property acquisitions and state arbitration between property rights and personal rights. The authors encourage justice thinkers to consider the diverse spectrum of libertarian thought when examining distributive, procedural and recognitional justice, and particularly how the ecological limits of natural resources present primary constraints on mainstream perspectives of absolute rights to property.

Similar to liberal theories of justice, Chapter 3, on cosmopolitan justice, focuses on both procedure and the substance of this approach to justice theorising. While cosmopolitanism long has idealised the universal inclusivity of humanity as subjects of a democratic community, scholars also note the difficulties of global procedural and distributive justice. As Skillington shows us in the chapter, despite the relative diversity within the cosmopolitan tradition regarding what constitutes 'just' relations, there nonetheless remains 'a focus on the moral, political, and legal status of "the citizen of the world" who in sharing planet Earth in common with others is bound by duties of hospitality and respect for global strangers' (Chapter 3). As she notes, the persistence of racism, war, gender inequities, and much more, challenge cosmopolitan ideals in critical ways. As such, approaches to cosmopolitan theorising within this tradition are highlighted which reimagine procedural and normative ways of realising what universal democratic justice may look like.

In Chapter 4, we see how feminist scholars have recentred who constitutes the traditional liberal subject of justice. While the question of why women were (and remain) excluded as rights-bearing subjects of justice has been central to feminist theorising, scholars more recently have begun to address concerns about who has the power to decide what is just and whether universal procedures for achieving justice can be realised given how different identity-based relationships dominate or oppress certain groups over others. Connected to procedural justice, the chapter addresses how feminist scholars understand the substantive forms that *injustice* takes, illustrating how lived experiences of oppression and domination occupy the normative core of justice theorising in this tradition. But as Mitchell shows us, feminist justice is as concerned with the forms as much as it is with the aspects of justice. Beyond the question of who constitutes a subject of justice and what the objective is when examining what is just about gender relations, Mitchell brings our attention to the ways in which feminist scholars highlight the domain or *where* of justice. Feminist scholars draw attention not only to women's exclusion from public participation but to the exclusion of the family unit as a legitimate scale through which just or unjust relations occur and can be theorised.

Along these lines, the approaches to justice theorising detailed in Chapter 5, covering the radical traditions of anarchist and Marxist thinking, as well as Critical Theory, focus heavily on the development of the normative substance of justice theorising, or its forms. But throughout this historical overview, we also see how this broad category of intellectual approaches places importance especially on identifying subjects and objects of justice and injustice, for instance, or its aspects. For example, Mitchell and Ohlsson show us how Marxist theorists, and critical theorists differently inspired by Marx, identify human emancipation as the object of justice, thereby establishing the human as the subject of justice; this is seen in anarchist desires to not be governed or by the socialist and Marxist emphasis on the way in which workers are alienated from their labour. Yet, as Mitchell and Ohlsson point out, for Marx, justice was only ever to be realised in the material relations of the world, not through ideal models of just relations. Although a Marxist conception of justice was never explicitly theorised by Marx himself, the Critical Theorists of the Frankfurt School developed upon such thinking, advancing critiques of justice theories, and eventually identifying procedures to achieve justice, such as with Jürgen Habermas' participatory goals around democratic consensus and Rainer Forst's right to justification. The chapter well illustrates how materialist and idealist approaches to theorising justice uneasily balance with one another, thereby providing opportunities for further engagement.

Throughout the chapters on postcolonialism (Chapter 6) as well as Indigenous approaches to justice (Chapter 7), explicit attention is given to the who or subjects of justice. Both chapters detail how the respective traditions developed in response to the exclusions of western thinking on these approaches to justice. For instance, postcolonial and Indigenous scholars both identify how non-western and Indigenous ways of knowing and being are often unaccepted as legitimate forms of theorising within mainstream justice theorising. Both chapters address how non-western knowledge production is remaindered if not entirely excluded from the collective canon of liberal approaches to justice theorising, which the authors identify as sources of *epistemic* injustice. Beyond recognising such injustices, the two chapters focus especially on the forms of justice. As Chapter 7 details, both the normative substance of Indigenous epistemologies and ontologies concerning what just relations with the world can and ought to be (processes of healing), as well as the procedural manner through which these ways of knowing and being can and ought to be realised (for example, respecting rights and treaties), highlight principles and instruments to maintain just relations within and outside of Indigenous communities. So too does postcolonial scholarship point to the need to restructure western intellectual traditions, particularly by analysing non-ideal theorisations of how and where injustices take place, by taking power asymmetries seriously, and by rethinking how conceptions of what is just ought to be addressed differently than solely through liberal mainstream discursive traditions.

The final chapter of this part of the volume, focusing on the Capabilities Approach (CA), describes the development of this popular approach. It details the positions of the approach's progenitors focusing on theorising inequality, a conceptual shift developed in opposition to the mainstream distributive theories, such as those described in Chapter 1. The overview also highlights how CA theorists identify procedures for measuring inequality and human well-being, in addition to how metrics for identifying social and environmental inequalities can be determined and addressed. Przybylinski and Sidortsov suggest that, while the CA is not a theory of justice in itself, the methodological flexibility of the approach positions it well to analyse situated inequalities within social, political, and environmental research.

To say that each of the chapters in Part I follow an identifiable pattern in the way that they detail the forms and aspects of each tradition would minimise the dynamism of justice theorising developed through each of these traditions. While some chapters herein may focus on certain forms or aspects over others, readers are encouraged to note the absences of other forms and aspects of justice in certain traditions. For, as we have already mentioned, each approach or tradition of justice does not speak for, nor define, justice in itself. By analysing matters of injustice through different ontological and epistemological positions, the chapters enable us to identify what is useful or problematic about justice theorising and thus how to situate justice within social science research. In general, taking the next step of situating knowledge of justice and injustice through various perspectives is the goal of applied fields of justice in the following chapters.

Part II: Applied justice theories

The chapters for Part II of the volume are positioned within more topical and empirically grounded areas of research. The applied approaches traditionally centre on specific objects of justice (for example, the environment and climate, energy systems, space and landscape, transitions, intergenerational relations). As such, each of the applied fields is conducive to analysing the aspects of justice. Yet, as the following chapters show, the aspects of justice foundational to these applied justice fields are sometimes just as concerned with addressing the principles and normative substance of justice theories as the normative traditions surveyed in Part I. The chapters thus also highlight how the initial concerns with topical matters within these research traditions have developed because of sustained engagement with political and normative theories of justice. In turn, we sometimes see how applied justice approaches have shaped normative justice theorising, by expanding how and why justice matters to the social, political and ecological.

Environmental justice (EJ) is one of the most robust justice concepts rooted in social justice, ecological justice and international environmental concerns. Chapter 9 details how this wide body of scholarship has emphasised the distributive, procedural, retributive and recognitional features of justice across various social categories, especially race, ethnicity and class. The movement for EJ connected to conceptualisations of EJ similarly seek to address the disproportionate impact of environmental harms and amenities on marginalised populations, as well as inequalities in decision-making processes and power structures. Wood-Donnelly argues that for this movement and set of discourses to be effective, EJ must consider both human and non-human aspects of environmental care and protection, and it must be applicable across geographical scales and timeframes. In doing so, it may better contribute to the improvement of human life as well as the protection of the environment as a whole.

Centring on how human induced climate change and the uneven effects and consequences of this are distributed, Chapter 10, on climate justice, details the ways in which justice researchers address responsibility for the actions taken and not taken to lessen the effects of climate change. It focuses on the procedural aspects of justice, detailing how institutions, processes and actors assume responsibility or hold culpability for climate injustices. Significantly, Skillington illustrates why an intergenerational perspective on justice requires us to think about how linkages between past, current and future climate scenarios are assessed and explored. Here, temporality is key. But alongside temporal relations, the chapter also describes the significance of scale as a concept within climate justice, as articulated through the cosmopolitan tradition, which often downplays the sovereignty of state territorial decision making, and instead focuses on our shared responsibility as humans for other humans, as well as for non-human beings.

The chapter on energy justice (Chapter 11) focuses primarily on how scholars in this nascent tradition have developed the concept through the forms of justice. The chapter identifies the ways in which energy has been conceptualised as a good to be produced, distributed and consumed, illustrating close theoretical engagement with the forms of justice. Yet, as Sidortsov and McCauley note, given the tradition's focus on energy systems as a whole, new contributions to this scholarship are beginning to stress the importance of, and need for, recognitional and restorative approaches to justice. In particular, the authors identify a need for those affected by changes in energy systems to be heard and legitimised participants in decision-making regarding these infrastructural challenges. In addition, the authors highlight the connectivity and fluidity of this body of justice scholarship, illustrating its overlapping, yet unique, foundations regarding energy as a primary object of justice analyses.

Chapter 12, on spatial justice, details the changing relationship of spatial thinking to justice theorising. While early spatial approaches to justice theorising developed with the analytical insights of liberal, distributive theories of justice, à la Rawls, the influence of more normative critiques of ideal justice theories began to reshape how justice mattered to spatial research several decades after Rawls. In the chapter, Przybylinski highlights how the thinking of feminist theorists such as Iris Marion Young and Nancy Fraser shaped how spatial theorists conceived of the concepts of justice and injustice, and thus with the role of space in situating injustices. This influence continues to frame spatial justice analyses today, particularly as seen through normative approaches using spatial relations as heuristics for framing injustices more broadly in society. With that said, no definitive theory of spatial justice has yet to be identified. As Przybylinski argues, one normative theory of spatial justice will likely not be useful for addressing injustices. Instead, he underscores that spatial thinking offers much by way of situating and identifying the conditions that lead to social, political and ecological injustices. As such, spatial justice offers a useful analytical framework through which to identify the domain of justice/injustice within social and spatial relations.

Similar to spatial justice, landscape justice (Chapter 13) is a research tradition primarily focusing on the humanly transformed environment. Here, issues of justice and injustice arise in the making, transformation and preservation of landscapes. The chapter details how the landscapes we live in, look at and take cultural value from also incorporate, reflect and advance just or unjust social relations. Landscape justice scholarship has long identified distinct objects (the production of landscape) and subjects (labourers) of justice, with a focus on the domain of injustices that are realised at different geographical scales. The extent of engagement with justice theorising from this scholarship, however, has largely been through mainstream, distributive theories of justice. As Mitchell notes, until recently, engagement with normative and philosophical theories of justice has largely been absent from the landscape justice tradition. However, Mitchell points to the recent normative turn in landscape justice research, engaging in especially feminist theories of recognition as well as with ethics, indicating how much space there is for landscape justice to develop more substantive analyses of justice which supplement its long-held interest in 'social justice' broadly conceived.

Chapter 14, on intergenerational justice, can be understood as a distinct tradition even as crucial aspects of it are intertwined with other justice traditions. For instance, intergenerational justice and the focus on justice for future generations are central to debates on climate justice in particular, but also environmental and energy justice as well as just transitions. What is distinctive about this tradition is its clear focus on time and temporality, as intergenerational justice highlights the short- and long-term future as well as the legacy of the past. Central issues discussed within this tradition are compound effects, the non-identity problem of future beings, and youth participation in decision-making processes.

The final chapter of the volume focuses generally on how to bring about a more just world that many of the other traditions envision. McCauley details how the just transition scholarship has yet to substantively engage with normative justice scholarship, particularly in identifying what the notion of a transition or transitions should be. Instead, issues central to the tradition have centred on workers' rights, in addition to issues related to energy production and distribution at local scales. Despite this, McCauley notes the potential presented by the lack of engagement with justice theorising. The under-theorisation of distributive and procedural forms of justice would usefully frame discourse, he notes. For instance, distributive and procedural forms of justice could help frame how workers are affected by shifting geographies of carbon and decarbonised development projects. Here, McCauley argues that the notion of vulnerability should become more central to transitions research, particularly as it relates to recognition and responsibility through the transition away from fossil fuel intensive developments, thus placing workers and families as subjects of injustice. McCauley thus points us to the ways in which the just transition scholarship remains open for engagement with justice theorising.

In the concluding chapter, we discuss overarching themes for Part I (the role of liberalism as a touchstone) and Part II (such as the connection between theoretical approaches and actually-existing circumstances). We also discuss how these two parts, and the traditions explored within them, are connected and where there is room for future research and potential for constructive dialogues. We offer a set of tables to help visualise and categorise the different traditions and hope that, despite their necessary reduction of detail, they provide readers with a pedagogical resource useful for drawing connections and indicating avenues for future research.

References

Fraser, N. (1997) *Justice Interruptus: Reflections on the 'Postsocialist' Condition.* New York: Routledge.

Harvey, D. (2009 [1973]) *Social Justice and the City.* Athens, GA: University of Georgia Press.

Jaggar, A. (2009) The philosophical challenges of global gender justice. *Philosophical Topics,* 37(2), 1–15.

Okin, S.M. (1989) *Justice, Gender, and the Family.* New York: Basic Books.

Rawls, J. (1999 [1971]) *A Theory of Justice,* revised edn. Cambridge, MA: Belknap Press.

PART I

Politico-Philosophical and Normative Traditions of Justice

1

Liberal Theories of Justice

Stephen Przybylinski

The predominant theories of justice come from the liberal tradition. Innumerable scholars and thinkers have scrutinised how the realisation of liberal values and instruments – equality, freedom and rights – may achieve justice or prevent injustices from happening. Given the historical depth and contemporary debate over liberal theories of justice, it is clear that much will be missed in any brief overview of liberal justice. To attempt some categorisation to the liberal tradition, the chapter outlines four of the most prominent theories of justice within it: social contract theories; deontological approaches; utilitarian approaches; and justice as fairness. Though these theories do not exhaust all liberal approaches to justice and are not mutually exclusive, the thinkers throughout the four traditions are indispensable to liberal theory historically. After introducing main ideas behind these traditions, the chapter identifies some of the main challenges presented by and within these dominant approaches. A particular emphasis is given towards the distributive aspects of justice as well as to how rights function as instruments to preserve liberal values. In ending, the chapter reflects upon who the subject of justice is assumed to be within liberalism and how demarcation of the liberal subject is justified.

Four liberal justice theories

Social contract tradition

The works of Thomas Hobbes, John Locke and Jean-Jacques Rousseau were central in the development of social contract theories of justice within liberalism. In general, social contact theories address how relationships of political authority come to be, by focusing on how sovereigns subject collectives and individuals to their authority. Authority is legitimated through a 'contract', to which all members of a given society adhere, which outlines the general moral and political obligations expected in that society.

Writing in the mid-17th century, Hobbes argued that without the presence of an authority in the form of a government, society would remain in a type of social chaos. Hobbes saw this chaotic state as the natural condition of humans, where everyone was 'against' everyone. In such a pre-civil state, Hobbes imagined that there would be no political sovereign present to resolve disputes or protect individuals' property. Without a sovereign to resolve disputes over rights, Hobbes argued, then the result would be that no individual action could be seen as 'unjust'. In this pre-civil state, 'notions of Right and Wrong, Justice and Injustice have there no place', he argued. For Hobbes, quite simply, 'where there is no common Power, there is no Law: where no Law, no Injustice' (1904, p 85). In other words, anything goes in the pre-civil state. Thus, to avoid the assumed chaos in this state of nature, Hobbes thought it was best that people consent to the authority of government by renouncing their 'rights of nature' so that they might be protected by civil rights essential to the establishment of society. Renouncing one's right to the state of nature and to establish civil society and political authority was, for Hobbes, the means by which individuals 'contracted' into a society.

Following closely from Hobbes, Locke was similarly concerned with avoiding the assumed social problems within the state of nature. Unlike Hobbes, Locke saw the state of nature not as a state of brutality per se, but one where individuals experience the truest extent of their natural liberties. More so than Hobbes, Locke (2016, ch III) saw the state of nature as one governed not by political authority but by the law of nature, a state of humanity whereby each individual (White male) is as naturally equal to every other (gender and racial inequality were not addressed politically or practically within Locke's thinking). Within the state of nature, then, individuals were endowed with *natural* liberty. Locke nonetheless recognised that individuals' natural liberties were vulnerable to the impositions of others within a state of nature. Thus, so too did Locke suggest that some social compact, regulated by a collectively agreed upon political authority, would be necessary to protect individuals' natural liberties (especially rights of property).

In the mid-18th century, Rousseau (2018) expanded on what legitimate political authority could look like and why a social contract is useful. Although Rousseau idealised the state of nature as one where individuals were most free, he saw the advance of conflict within modern society as the very reason that a social compact was necessary. Leaving the state of nature could only be legitimate, he argued, when a 'sovereign', such as a democracy, was given its consent by the general will of the people. The benefit for individuals in 'leaving' the state of nature, he argued, is that they gain 'civil liberty'

(Rousseau, 2018, p 26), a new liberty that was needed as individuals' natural liberties were no longer ensured by individuals' subservience to others. Civil liberties would need to be protected and governed according to the general will of people. It would be only through this democratic commitment to shared power that individuals would be equal 'by convention and legal right' (Rousseau, 2018, p 31). Unlike other state of nature theorists, therefore, Rousseau's argument for government advanced the notion of publicly held power as key to a just political society.

Justice within the contractarian tradition is advanced when individuals consent to a relationship with some type of sovereign. The social contract is the realm through which individuals realise a sense of social justice not only for themselves but for others, by maintaining a political state where individual and societal liberties are preserved. For Rousseau, because justice was nearly an intuition, all individuals were endowed with a 'sense' of justice which made them capable of self-governance. Hobbes put less stock in the legitimacy of democratic practice than Rousseau. More simply, justice for Hobbes was realised when individuals simply respect the rights of contract, or what he calls a 'covenant' (Hobbes, 1904, p 97). Justice within the contractarian tradition is somewhat straightforward in this sense. Society is just when a naturally free people voluntarily consent to be governed and respect the laws of contract upheld by political authority. The notion of a contract has remained central to liberal doctrine to this day, as we will see in the section on justice as fairness.

Deontology

Deontology is a normative theory that holds that there are morally permitted or forbidden choices for individuals to make. Deontological theories are used to 'guide and assess our choices of what we ought to do ... in contrast to those that guide and assess what kind of person we are and should be' (Alexander and Moore, 2020). An absolutist approach to deontology suggests that there are 'correct' choices to be made in so much as they conform to moral norms. In this way, deontologists prioritise the 'right' over that of the 'good' (Alexander and Moore, 2020). That is, deontologists are more concerned that morally correct actions are identified and followed than with theories that seek to develop differently experienced senses of good or value by following whatever actions are necessary to realise that good.

A central contributor to deontological theory has been Immanuel Kant. Kant was concerned with respecting individuals as free, autonomous, rational agents with dignity. From Kant's perspective, it is wrong to impede on another individual's ability to pursue their own ends. Of course, individuals do impede on the agency of others and restrict certain freedoms as a result of political relations within society. This makes the ethical principle to never impose on the autonomy of others difficult to follow. Kant's thinking here is thus somewhat challenging to balance with his political thinking on justice.

Kant's universal theory of justice states simply that one's choices are just if they 'act externally in such a way that the free use of your will is compatible with the freedom of everyone according to a universal law' (1965, p 35). In contrast, self-serving actions are not generalisable as universally applicable rules that will hold for all other moral agents. One thus respects the humanity in others by acting in accordance with rules that hold for everyone.

For Kant, political obligations follow from our ethical and moral obligations. As such, Kant's notion of justice is a matter of obliging one's political duties that follow from moral duties. Our moral duties are a matter of ethics and we cannot compel others to be virtuous. However, Kant suggests that we can compel others to act justly as a *juridical* duty to others. Under Kant's universal law, 'when we do our juridical duty, we do what others can rightfully demand that we do' (Aune, 2016, p 133). So too does this mean that 'strict justice relies ... on the principle of possibility of external coercion that is compatible with the freedom of everyone in accordance with universal laws' (Kant, 1965, p 37). That is, the power of political authority is justified only to ensure that everyone's freedom is not impinged upon when upholding the law.

Justice under Kant's universal law is a rights-based (and duty-based) relationship. Rights are the means to protect individuals as free and moral agents. Given this, Kantian justice espouses a deontological liberalism, one in which the 'purpose of justice is not to legislate positive law that embodies ethical conduct in the sense of the actions that rational beings "legislate" their own behavior, but to provide a framework of predictable external security of body and property, within which rational persons may exercise their own autonomy' (Campbell, 2010, p 72). From this perspective, an individual's 'predictable external security' is protected through universal rights which allow individuals to remain unencumbered in pursuing their own vision of a good life. By 'simply' acting in accordance with what is 'right' (and through these actions, a corresponding duty to others), the rational individual of Kantian justice is thereby acting justly. Crucially, deontological ethics has shaped the idealised liberal individual by affirming that political rights protect individuals for the purpose of maintaining their autonomy.

Utilitarianism: justice as a means of advancing social utility

Contra deontological approaches to justice, utilitarianism is a perspective generally holding that morally right actions are those that produce the greatest amount of good. As one of the preeminent utilitarians, Jeremy Bentham helped establish the idea that happiness is the primary ethical value or 'good' around which society ought to organise itself (Bentham, 1907). As a hedonistic theory, Bentham's utilitarianism sees that 'only pleasure is intrinsically good and only pain intrinsically evil' (Hurka, 2007, p 359). In this way, Bentham's principle of utility 'approves or disapproves of *every* action according to the tendency it appears to have to increase or lessen – i.e. to promote or oppose – the happiness of the person or group whose interest is in question' (Bentham, 1907, p 2, emphasis in original). To maximise happiness and avoid pain is to maximise social utility.

Although Bentham never developed a utilitarian theory of *justice*, others such as J.S. Mill advanced a theory of justice premised on principles of utility. Mill suggests utility be measured not through an *individual's* happiness but through the 'greatest amount of happiness altogether' (1951, p 14). This collective moral position provides the key to Mill's theory of justice. Justice, Mill states, 'supposes two things: a rule of conduct and a sentiment which sanctions the rule. The first must be supposed common to all mankind and intended for their good. The other (the sentiment) is a desire that punishment may be suffered by those who infringe the rule' (Mill, 1951, p 65). Here justice is found through 'objective' moral rules to which society must adhere. 'Justice is a name for certain classes of moral rules, which concern the essentials of human well-being more nearly, and are therefore of more absolute obligation, than any other rules for the guidance of life' (Mill, 1951, p 73). From this perspective, our societal obligation to not increase pain or unhappiness for others while maximising happiness will be fulfilled by respecting moral rules.

Mill may be seen as a 'rule utilitarian' in that justice is realised by respecting the rights of individuals. To that end, Henry Sidgwick more closely tied law into the concept of utilitarian justice. Sidgwick (1907) found law to be the concrete means through which to balance the broad differences between 'conservative' and 'ideal' justice. 'The laws in which Justice is or ought to be realised, are laws which distribute and allot to individuals either objects of desire, liberties and privileges, or burdens and restraints, or even pains as such' (Sidgwick, 1907, p 266). This is conservative justice. Thus, upholding contracts is essential to conservative justice (Sidgwick, 1907, p 270). But Sidgwick also recognises there is ideal justice as well, where freedom is seen as 'the ultimate end and standard of right social relations' as well as one founded on 'requiting desert' à la universal receipt of material goods (1907, p 294). Because it is impossible to 'obtain clear premises for a reasoned method of determining exactly different amounts of Good Desert', Sidgwick argued that we must give up on 'construction of an ideally just social order' (1907, pp 290–291). Unlike contract theories of justice, and anticipating the Rawlsian idealised version of a basic (and just) social structure, Sidgwick argues that inequality will ever be part of social life. As such, Sidgwick reinforces the utilitarian conception of justice, being primarily a duty of upholding social contracts through legal means, as the best means towards maintaining the greatest good overall.

Justice as fairness

The most well known theory of liberal justice is that of John Rawls' 'justice as fairness'. In *A Theory of Justice*, Rawls (1999) devises a contractarian notion of justice meant not to define what justice *is* but to establish moral principles that would be just for the basic structure of society. It is a contractarian theory in that Rawls creates a hypothetical situation for a just social order, what he terms the 'original position', by which individuals agree to abide. The original position has two main principles: that each individual has an equal right to (1) the most extensive scheme of equal basic liberties and (2) the fair distribution of social and economic goods as well as positions and offices. These two principles must be followed, Rawls suggests, for 'they provide a way of assigning rights and duties in the basic institutions of society and they define the appropriate distribution of the benefits and burdens of social cooperation' (1999, p 4). Thus, Rawls' conception of justice is contractarian and deontological, in that these two principles prioritise adherence to what is *right* as the proper move towards realising justice rather than conceptualising a notion of justice that seeks to define what is *good*. The premise of the original position is that it would ideally preserve the autonomy of the rational, liberal subject while at the same time develop a means of establishing an equitable distribution of primary goods among society.

Rawls' theory of justice is said to be 'fair' in that the principles are agreed to by free and rational persons who would further their own individual interests within an initial position of equality (1999, p 10). To arrive at this fairness, Rawls conceives of a 'veil of ignorance' within the original position in order to establish an equal grounding for individuals. Under the veil of ignorance, individuals do not know certain kinds of facts about themselves. No one knows their place in society, their class position or social status; their fortune in the distribution of natural assets and abilities; their own conception of the good; the particular circumstances of their own society, that is the political or economic situation; and they have no indication to which generation they belong. As far as possible, then, 'the only particular facts which the parties know is that their society is subject to the circumstances of justice and whatever this implies' (Rawls, 1999, p 119). Rawls believes that the veil of ignorance avoids the shortcomings of other theories of justice, in particular, utilitarianism, which he argues promotes the idea that everyone must be benevolent for there to be social justice. Instead, 'the combination of mutual disinterest and the veil of ignorance achieves much the same purpose as benevolence', he argues. 'For this combination of conditions forces each person in the original position to take the good of others into account. In justice as fairness, then, the effects of good will are brought about by several conditions working jointly' (Rawls, 1999, pp 128–129). The function of the veil of ignorance, therefore, is to maintain not only the priority of liberty protections essential to liberalism, but also to reinforce the notion of the disinterested individual as the liberal subject.

Rawls' theory is a paradigmatic liberal theory because it prioritises the protection of liberty rights to regulate the basic structure of society. At the same time, Rawls promotes egalitarian outcomes for all individuals. Most notably he does this by developing the 'difference principle', which is founded on liberal ideas of equality of opportunity. As detailed in the next section, the difference principle attempts a proper balance between economic distribution and political rights. Since Rawls, debates over just this balance have predominated in theories of justice. As such, the following section expands on the definitive role of rights discourse and of distributive theories of liberal justice before attending to some key critiques of them.

Debates on redistribution and rights

(Re)Distribution

Any theory of justice is about (mal)distribution of some thing or process in some respect. But liberal theories of justice have long centred on the just distribution of two things in particular: material goods and rights. Rawls' justice as fairness is exemplary for attempting to balance both, but particularly for the manner in which it works through equitable distribution. Rawls (1999) uses what he terms the 'difference principle' to explain how an equitable distribution of primary goods would be arrived at fairly in his hypothetical just society. The difference principle states that 'social and economic inequalities are to be arranged so that they are both (a) to the greatest expected benefit of the least advantaged and (b) attached to offices and positions open to all under conditions of fair equality of opportunity' (Rawls, 1999, p 72). Rawls' theory is egalitarian in the sense that it requires that all individuals are to receive the same amount of primary goods – income, access to political office, and so on. If individuals should agree upon something other than equal distribution, then the difference must benefit the least well off. The point for Rawls was to devise a way to provide individuals with equality of opportunity through distributive means.

Some within the liberal tradition have argued, however, that Rawls' difference principle would not establish equality among individuals. Ronald Dworkin (2000), for instance, argued that because Rawls does not consider individual talents and abilities under the veil of ignorance, that natural disparities will arise between individuals after an initial equal distribution of goods. Rather than Rawls' equality of opportunity, Dworkin argues for an 'equality of resources' to try and balance equality in resources over time. Dworkin's notion of equality of resources 'supposes that the resources devoted to each person's *life* should be equal' (2000, p 70, emphasis added). For Dworkin, what is needed to ensure this life equality is a market device (an auction) whereby individuals can relate to each other about what they value. 'The true measure of the social resources devoted to the life of one person is fixed' by this auction, 'by asking how important that resource is for others' (Dworkin, 2000, p 70). Dworkin's resources theory 'supposes equality defines a relation among citizens that is individualized for each, and therefore can be seen to set entitlements as much from the point of view of each person as from that of anyone else in the community' (2000, p 114). In Dworkin's auction, then, individuals figure out what they value in their lives and resources are distributed accordingly based on this information. When natural inequities arise in resources because of differing talents, Dworkin suggests a tax to enable ongoing redistribution to try to maintain material equality among people.

More so than Rawls, Dworkin's equality of resources attempts to find a just distribution *between* individuals rather than among a group of individuals collectively. In Dworkin's theory, individuals are equal because their differences are maintained, differences in talents and abilities which Rawls' hypothetical contract removes through the veil of ignorance. Thus, for Dworkin, 'equality is in principle a matter of individual right rather than one of group position' (2000, p 114). This perspective promotes a more individualised distribution and protection of material goods, a perspective more inclined to free-market principles of distribution than that of Rawls' equality of opportunity.

Such distributive theories have been criticised from within and outside of the liberal tradition. Closely aligned to the principles of liberalism, 'capabilities' theorists argue that we ought not to focus simply on whether there is an equal distribution of primary goods but on how individuals can or cannot make use of those primary goods. Amartya Sen, the originator of the Capabilities Approach, has critiqued distributive theories for focusing too much on the *means* of living. Instead, Sen argues we ought to be evaluating the actual *opportunities* a person has to do the things they value doing (2009, p 253). While the Capabilities Approach recognises that material equality is foundational in maintaining an equitable society, many reject that it is possible to identify a perfectly just distribution, and seek instead to evaluate matters of injustice based on individual well-being (see Chapter 8).

Another critique of the liberal emphasis on distributive justice comes from right-leaning libertarianism. Absolutist right-leaning libertarians reject that there should be distribution of primary goods by any means other than voluntary exchange through the free market. Here, inequalities in society are seen as naturally occurring and acceptable. In defending this position, Nozick (1974) suggests that inequalities in 'holdings' (primary goods) are *just* when an individual has 'fairly' acquired a resource, which individual procure by cultivating 'unheld things' with their labour or from a transfer through free market exchange (pp 150–153). From this perspective, once a person possesses a resource, they then hold inviolable rights to those resources. To redistribute an individuals' holdings is thus unjust in that it violates their rights of property, what Nozick sees as an unfair taking of an individual's holdings.

Libertarian notions of justice recognise no form of redistribution outside of free market exchange. Under Nozick's entitlement theory, 'a distribution is just if everyone is entitled to the holdings they possess under the distribution' (1974, p 151). From this view, there should be *no* principle of redistribution through which individuals equate their resources, such as that found in Rawls' equality of opportunity or Dworkin's equality of resources. Instead, to adjust for inequalities in social goods, individuals make the voluntary choice to offer charity to others (Nozick, 1974, p 235). The common liberal redistribution process of taxation, along with principles devising equality in resources, cannot be just from the libertarian perspective in that such a redistribution unfairly takes from an individual what is 'rightly' theirs. For, as Friedrich Hayek (1998, p 64) argues, iniquitous outcomes from the market that were not initially intended are still justified, and to 'demand justice' for these inequities by singling out people would be unjust.

A final set of debates over the justness of distributive theories was advanced during the post-structural turn in social sciences a few decades after Rawls' justice as fairness. Feminist scholars in particular illustrated the need for more substantive considerations in theories of justice that moved beyond a narrow focus on ideal equal distributions. Iris Marion Young (1990), for instance, argued that examining the distribution of primary goods alone was insufficient to understand what is just and, therefore, that any conception of justice must also address the processes by which individual and group liberties are actively violated. Distributive theories, Young argued, conceptualise social justice 'primarily in terms of end-state patterns, rather than focusing on social processes' themselves (1990, p 25). Overemphasising distribution, she suggests, ignores how social structures constrain or deny the self-determination of individuals or groups.

Similar to Young, Nancy Fraser (1997) argued that a notion of justice which only attends to socioeconomic matters of redistribution misses the related social and cultural injustices extant within liberal society. Her notion of justice as 'recognition' underscores Young's argument that cultural misrecognition and the domination of specific groups of people are left unaddressed in theories of distributive justice. Justice for Fraser necessitates both economic redistribution *and* a politics of recognition (and later participation) which seeks equality for traditionally dominated social groups. The more substantive contributions of these critiques broadened the scope of justice beyond a matter of imagining ideal distributions among society.

Rights

Liberalism ideally protects a basic set of liberties for individuals: freedom of conscience, freedom of thought and discussion, freedom of association, freedom of choice in occupation and to participate within the free market, and the protection of property, among others (Freeman, 2018, p 64). These political and economic freedoms are protected through personal *rights*. And while rights discourse is not limited to theories of justice alone, rights nonetheless remain fundamental to any liberal conceptualisation of justice.

Liberal rights are not one and the same. Rights can be understood as relationships, moral claims and legal instruments, among other rights types. When it comes to legal rights, it is common to make the distinction between 'negative' and 'positive' liberty rights, for example. A negative liberty constitutes a warding off of, or 'freedom *from*', interference, such as from an overreaching state towards an individual, while a positive liberty suggests a 'freedom *to*' act, a possibility to control one's life (Berlin, 2002, p 178, emphasis added). Negative liberties are associated with what are called first-generation rights, or the civil and political rights of citizenship that 'require only that we and our governments refrain from various acts of tyranny and oppression' (Waldron, 1993, p 24). Negative rights are contrasted with second-generation rights or rights to socioeconomic goods, which 'correlate to positive duties of assistance' (Waldron, 1993, p 24). Positive rights require the state to provide some good, such as welfare benefits, to a polity in general. Despite their relative limitations, positive rights are critical for helping maintain basic levels of economic security for individuals. Both first- and second-generation rights are enshrined into international law, such as through the Universal Declaration of Human Rights, the International Covenants on Civil and Political as well as Economic, Social, and Cultural Rights (United Nations, 2022).

Negative rights are by far the most common form of rights codified within liberal law. For, they tend to emphasise the scope of power an individual enjoys by representing the autonomy of the citizen subject within liberalism (Nedelsky, 1993). To confer a liberty right on someone, Norberto Bobbio states:

> is to recognize that the individual in question has the capacity to act or not to act just as he pleases, and also the power to resist, availing himself in the last instance of the use of force (his own or others'), against whoever may transgress that right: so that potential transgressors have in turn a duty (or obligation) to abstain from any action which might interfere in any way with this capacity to act or not to act. (Bobbio, 2005, p 5)

Liberty rights are legal protections which secure for individuals a basic set of freedoms that are thought to enable a good life.

Membership in liberal political communities affords individuals these basic protections. But many have been excluded from liberal polities, leaving certain groups unprotected in basic human freedoms. As nation states have traditionally ruled through particular ethno-nationalist criteria, for example, so have cultural minorities been excluded or deeply disadvantaged by the sociopolitical and economic regulations of a given state. The delegitimisation of Indigenous sovereignty by nation states in many places around the world represents just this inequity (Moreton-Robinson, 2005).

Yet, rights can be critical for bridging disparities in power as well. Will Kymlicka (1995, p 6) has argued for specially recognised group rights (sometimes called 'third-generation' rights) for cultural minorities as a means of addressing the sociopolitical and economic disparities between minority groups within nation states. For marginalised groups that hold citizenship within a nation state, but have been and continue to be oppressed through legacies of hierarchical racial or ethnic relations, rights can also be a necessary step towards balancing historical inequities. Against arguments that rights lack utility in that they obscure true need, Patricia Williams has argued for the necessity of rights (positive and negative) for African Americans, for example. Rights, for Williams (1992, p 152, emphasis in original), provide 'a political mechanism that can confront the *denial* of need' for oppressed groups, the denial of need oft brought by those in power. Rights can elevate the status of marginalised or oppressed groups while forcing those holding power into relations of respect for the marginalised.

While rights in themselves may not bring forth a sense of 'justice', rights offer an extant means towards realising equality among individuals and groups, one foundational to liberalism as a political and moral project. As such, any liberal theory of justice must acknowledge what role rights play in collapsing inequitable relations of power and protecting individuals from harms by overreaching individuals and government.

Criticisms of liberal justice: *who* is a subject of justice within liberalism?

Critics of liberal justice have shown that, beyond matters of distribution and rights-protections, the ways in which justice theories have idealised the liberal subject have been all but inclusive. If the individual is the scale on which liberalism is centred, just *who* is able to make claims for justice has often been presupposed within justice theories. One scholar sensitive to liberalism's ideals yet critical of how the development of liberalism has been exclusionary of others is the late Charles Mills. Mills (2017) argues that liberal theories of justice do not specifically address race and the injustices of White supremacy within dominant liberal discourse. He notes, for instance, that social contract theories – like Rawls' – are 'better thought of as an exclusionary agreement among whites to create racial polities rather than as a modeling of the origin of colorless, egalitarian, and inclusive socio-political systems' (Mills, 2017, p 140). A second critique follows from this first claim: distributive theories of justice do not identify a means of rectifying historical racial injustices, in that they idealise the perfectly just society that *could* be. In particular, Mills notes how Rawls offers no theory of rectificatory justice, or a theory attempting to correct for injustices through the non-ideal or 'world as it is', thereby ignoring the oppression that ought to be central to such an ideal theory of justice.

Other critics of liberal justice similarly stress why differences like race, ethnicity or gender must matter to conceptualisations of justice. For some, the impartiality of deontological approaches, for instance, can be blind to the differences in value between liberal subjects. Universal, rights-based approaches can enable 'those exercising political and administrative power – principally men – to generalize from their own experiences and to neglect those that are foreign to them' (Campbell, 2010, p 189). For Iris Marion Young, the implications of impartiality are problematic. To bring into examination gender and other social differences is necessary to understand injustice, Young argues, as justice 'requires not the melting away of differences, but institutions that promote reproduction of an aspect of group differences without oppression' (1990, p 47).

Others still suggest that real equality among society cannot be achieved within the bounds of liberalism. Marx, for instance, was suspect of liberal rights, arguing that at best, they only provide the 'final form of human emancipation *within* the framework of the prevailing social order' (Marx, 1978, p 35, emphasis in original). As he notes, liberalism is a social-political order dominated by the interests of the bourgeois ruling class. As such, the function of rights within liberalism reinforce and protect the needs of liberal economic organisation and therefore structure how and who the liberal state may protect in disputes over rights.

Just as liberal theories have remained prevalent within justice discourse, so too have the critiques of liberalism remained substantial. Critiques of liberal justice advance new understandings of what may constitute justice or injustice, identifying who may be excluded from the protections of liberal citizenship or how ideal justice theories focused on distribution may be less useful for addressing *why* certain groups do not have equality in goods broadly understood. They illustrate that finding solutions to inequalities cannot be premised in ideal theories of justice alone, certainly without also paying attention to how power is experienced within political society. Significantly, they help illustrate how the idealised autonomy of the liberal individual is not equally experienced by, and protected for, all people.

Making a connection to the social sciences

As this chapter has shown, liberal theories are ever evolving and remain a focal point in justice theorising more generally. The continued relevance of liberal themes addressed here is reflected in much discourse today. While the uneven effects in distribution of socioeconomic goods clearly threatens individuals' economic stability within and between liberal states today, so too do rights and obligations for rights embed themselves in dialogues about who gets to make decisions about what and where. To take one example, the equity in distribution of primary goods among Indigenous communities within liberal states across the world pertains not only to disparities of individual incomes and economic security. The right of minority groups to not be burdened by direct or indirect environmental effects of new or past developments, to have the right to genuine participation and capacity to affect decisions made about developments within one's own community, and to have legitimate mechanisms by which to hold those accountable who transgress sociopolitical agreements, rely upon protections of rights (or lack thereof) and of autonomy as subjects within liberal polities.

In that liberty is paramount for liberal ideology, so then is liberty central to justice within liberalism. What this chapter has stressed is that rights play a pivotal role in securing basic political freedoms for individuals. While justice is not simply the protection of rights outright, critical dialogue around how rights do and do not benefit liberal subjects helps to assess more specifically what we understand to be just or unjust in the world. An individuals' right to compensation for the burdens of energy development on their livelihood, for instance, does not correlate to a right to have one's livelihood damaged by the environmental externalities of oil extraction at the same time. It does not necessarily follow that just compensation for ecological damages to a local community also enables just social or environmental relations for individuals in that community with each other, society more broadly, or with the environment. At the same time, group rights for politically and economically marginalised groups may offer the ability to affect changes that collectively benefit the communities in which developments take place. 'Positive' rights may offer a key means to not just economic goods for under-resourced communities, but perhaps even the ability to hold individuals and groups accountable for inequities arising from developments.

What liberal theories of justice and its critiques help us to see is that with changing understandings of who is included as a liberal subject and who has a legitimate ability to enact change within liberal polities, theories of justice may better reflect the ideals of liberalism. This means that liberal theory must continue to reckon with how it defines its boundaries, forcing it to reflexively adapt to new justifications about how liberal ideals are realised and for whom.

References

Alexander, L. and Moore, M. (2020) Deontological ethics. In E.N. Zalta (ed), *The Stanford Encyclopedia of Philosophy*. https://plato.stanford.edu/archives/win2020/entries/ethics-deontological/

Aune, B. (2016) *Kant's Theory of Morals*. Princeton: Princeton University Press.

Bentham, J. (1907) *An Introduction to the Principles of Morals and Legislation*. Oxford, UK: Clarendon Press.

Berlin, I. (2002) Two concepts of liberty. In H. Hardy (ed), *Liberty: Incorporating Four Essays on Liberty*. Oxford: Oxford University Press, pp 167–218.

Bobbio, N. (2005) *Liberalism and Democracy*. London: Verso.

Campbell, T. (2010) *Justice*, 3rd edn. Camden: Palgrave Macmillan.

Dworkin, R. (2000) *Sovereign Virtue: The Theory and Practice of Equality*. Cambridge, MA: Harvard University Press.

Fraser, N. (1997) *Justice Interruptus: Critical Reflections on the 'Postsocialist' Condition*. Abingdon, UK: Routledge.

Freeman, S. (2018) *Liberalism and Distributive Justice*. Oxford, UK: Oxford University Press.

Hayek, F. (1998) *Law, Legislation, and Liberty: A New Statement of the Liberal Principles of Justice and Political Economy*. Abingdon, UK: Routledge.

Hobbes, T. (1904) *Leviathan: Or, The Matter, Form & Power of a Commonwealth, Ecclesiastical and Civil*. London: C.J. Clay and Sons.

Hurka, T. (2007) Value theory. In D. Copp (ed), *The Oxford Handbook of Ethical Theory*. Oxford, UK: Oxford University Press, pp 358–380.

Kant, I. (1965) *Metaphysical Elements of Justice*. Indianapolis, IN: Bobbs-Merrill Educational Publishing.

Kymlicka, W. (1995) *Multicultural Citizenship: A Liberal Theory of Minority Rights*. Oxford, UK: Oxford University Press.

Locke, J. (2016) *The Second Treatise of Government*. Indianapolis, IN: Hackett.

Marx, K. (1978) On the Jewish question. In R. Tucker (ed) *The Marx-Engels Reader*. New York: W.W. Norton & Company, p 35.

Mill, J.S. (1951) *Utilitarianism, Liberty, and Representative Government*. Boston, MA: E.P. Dutton and Company.

Mills, C. (2017) *Black Rights/White Wrongs: The Critique of Racial Liberalism*. Oxford, UK: Oxford University Press.

Moreton-Robinson, A. (2005) Patriarchal whiteness, self-determination, and Indigenous women: The invisibility of structural privilege and the visibility of oppression. In B. Hocking (ed), *Unfinished Constitutional Business: Re-thinking Indigenous Self-determination*. Canberra, Australia: Aboriginal Studies Press, pp 61–73.

Nedelsky, J. (1993) Rights as relationships. *Review of Constitutional Studies*, 1(1), 1–26.

Nozick, R. (1974) *Anarchy, State, and Utopia*. New York: Basic Books.

Rawls, J. (1999) *A Theory of Justice*, revised edn. Cambridge, MA: Belknap Press.

Rousseau, J.J. (2018) *The Social Contract*. London: Arcturus Publishing.

Sen, A. (2009) *The Idea of Justice*. Cambridge, MA: Harvard University Press.

Sidgwick, H. (1907) *The Methods of Ethics*, 7th edn. New York: Macmillan and Co.

United Nations (UN) (2022) *International Covenant on Civil and Political Rights*. https://www.ohchr.org/en/instruments-mechanisms/instruments/international-covenant-civil-and-political-rights

Waldron, J. (1993) *Liberal Rights: Collected Papers 1981–1991*. Cambridge, UK: Cambridge University Press.

Williams, P. (1992) *The Alchemy of Race and Rights: Diary of a Law Professor*. Cambridge, MA: Harvard University Press.

Young, I.M. (1990) *Justice and the Politics of Difference*. Princeton, NJ: Princeton University Press.

2

Libertarian Theories of Justice

Corine Wood-Donnelly, Darren McCauley and Stephen Przybylinski

Introduction

A libertarian theory of justice, so called, must be a system of justice that is based on the ideals of libertarianism. In general, libertarian justice can always be reduced to the precept of whether an action violates the principle rights of liberty. The theorist Brennan remarked that 'Libertarianism is a demanding doctrine – it demands that we mind our own business, even though most of us would rather not' (Brennan, 2012, p 3). While libertarian justice has much overlap with liberal theories of justice, mainstream libertarianism holds a much narrower understanding of justice. The rights to property protection are essential to liberty protections. Any theory of libertarian justice evaluates what is just by assessing whether property is protected and distributed according to how individual rights have been respected. In this sense, libertarian justice is more procedurally transparent than that of liberal justice.

While strict in definition, libertarian conceptions of justice are notably complicated in their outcomes. As this chapter will explicate throughout the following sections, the adherence to a conception of justice based on strict protection of individual rights alone flies in the face of common understandings of justice as a means towards equality. Whereas liberal conceptions of equality require both the protection of individual rights and a means of creating equitable outcomes in the distribution of primary goods, mainstream libertarianism requires only equal protection of individual rights. As the chapter will detail, this difference in how equality is preserved matters greatly. After detailing the essence of rights from a libertarian perspective, the chapter explains what libertarians define as just procedure and how this process shapes distribution. The section following that traces the historical development of libertarianism to show how the predominant understanding of libertarianism and libertarian justice does not constitute the entirety of libertarian thought, suggesting that there is room for critique within libertarianism as commonly understood.

Throughout the chapter, we distinguish between 'mainstream' and 'left' libertarianism. We understand mainstream libertarianism to be the dominant understanding of libertarianism today, which is an economically conservative and thus 'right' leaning libertarian perspective. Although dominated by the mainstream, left-leaning libertarian positions have been around for much longer. Nonetheless, libertarianism particularly in Anglo-political theory has come to be associated with the political right. In the places throughout the text where libertarianism is used by itself, it refers to the broad tradition of libertarianism, incorporating both right- and left-leaning perspectives.

The basis of libertarian justice is rooted in natural rights. Natural rights doctrine understands that all individuals hold inherent rights as humans, which exist outside of and before rights deriving from the laws of political society. While rooted within ecclesiastical foundations, after the Enlightenment period, natural rights have come to be understood through the means of reason and rationality. From a natural rights perspective, then, rights are moral and enforceable claims to respect (James, 2003). As Locke (2016 [1689]) argued in the late 17th century, all men (women were property themselves) have natural rights to life, liberty and estate. To have a natural right to these things is to enjoy natural liberties that exist for humans outside of political and civic society.

Locke, among others, noted, however, that to best preserve natural rights, political authority is needed to protect natural liberties. For, within a state where humans are naturally free – the state of nature – the potential for anarchy would surely impose on people's liberty. Describing such limits, Hobbes warned that in the state of nature:

> [E]very man has a Right to every thing; even to one anothers body. And therefore, as long as this naturall Right of every man to every thing endureth, there can be no security to any man (how strong or wise soever he be) of living out the time, which Nature ordinarily alloweth men to live. (Hobbes, 1904, p 87)

To ensure that humans would be able to preserve themselves, or their 'nature', a political sovereign was needed to uphold natural rights in society. In other words, a state was needed to enforce the natural liberties of individuals.

When rights are enforced through a state, rights are embedded in or adhere to the individual, rather than in society per se. The libertarian and liberal traditions share in common the need to protect individual rights, particularly 'liberty' rights. To attribute a liberty right to someone:

is to recognize that the individual in question has the capacity to act or not to act just as he pleases, and also the power to resist, availing himself in the last instance of the use of force (his own or others'), against whoever may transgress that right: so that potential transgressors have in turn a duty (or obligation) to abstain from any action which might interfere in any way with this capacity to act or not to act. (Bobbio, 2005, p 5)

Liberty rights thus are legal protections which secure for individuals a basic set of freedoms fundamental for leading a good life.

Liberty rights are not all the same and it is common to make the distinction between 'negative' and 'positive' liberty rights. A negative liberty constitutes a warding off of, or 'freedom from', interference, while a positive liberty suggests a 'freedom to' act (Berlin, 2002, p 178). Negative liberties are associated with what Jeremy Waldron (1993) calls first-generation rights, or the civil and political rights of citizenship that 'require only that we and our governments refrain from various acts of tyranny and oppression' (p 24). Negative rights are contrasted with second-generation rights, or rights to socioeconomic goods, which 'correlate to positive duties of assistance' (Waldron, 1993, p 24). Such 'positive' rights require assistance from the state to provide some good, such as welfare benefits, to society in general. Positive rights are critical for helping maintain basic levels of economic security for individuals and are understood to enhance equality rather than protect liberty.

Libertarianism and liberalism diverge from one another in regard to which type of liberty rights are justifiable. In libertarian thought, preserving liberty means strictly enforcing negative rights; the right to be free from coercion, force and the effects of power. To realise positive liberties requires an active state and as such is rejected from a libertarian viewpoint. For this reason, libertarianism is said to be based upon the 'principle of maximum liberty' (Kaufmann et al, 2018). The strict adherence to negative rights is found within the opening lines of the preeminent libertarian text of the late 20th century, Robert Nozick's *Anarchy, State, and Utopia*. Nozick opens the book with the following: 'individuals have rights, and there are things no person or group may do to them (without violating their rights). So strong and far-reaching are these rights that they raise the question of what, if anything, the state and its officials may do' (1974, p ix). To violate individuals' rights, from this perspective, is to treat individuals not as ends in themselves but as a means towards something else, violating the natural liberty of the individual. What follows from this is that the state can be no more than 'minimal' in the lives of individuals, organised solely to prevent fraud, theft and to uphold contracts.

From the libertarian view, therefore, respect for individual liberty is the central requirement of justice (Brennan et al, 2018). Essential for respecting individual liberty within libertarianism is the upholding of rights of contract. Property rights, and more specifically ownership, are key conceptual principles that underpin libertarianism and its justice framework. Again, Locke's arguments for a private right to property are instrumental to the libertarian emphasis on protecting property. Locke (2016 [1689]) suggested that by investing one's own labour into a resource, property became properly owned by that individual. In other words, 'property' was invested within the individual and, as such, individuals were self-owning. The ownership principle of libertarianism, built on a moral basis of natural rights and individual liberty, is the starting point for examining libertarian justice. In what follows, we delve further into how these core principles shape the libertarian theory of justice.

Overview of main ideas and scholars within the tradition

The primary foundation of libertarianism is rooted in individual liberty with the individual viewed as an owner of themselves. The emphasis on the self-ownership of the individual facilitates private ownership as a product of one's labour. The idea of a just distribution, then, is fairly straightforward for libertarians. As Nozick (1974, p 151) argued, a 'complete principle of distributive justice would say simply that a distribution is just if everyone is entitled to the holdings they possess under the distribution'. Nozick's 'entitlement' theory is therefore representative of a mainstream (right-leaning) libertarian theory of distributive justice.

Nozick's entitlement theory argues that distributions of goods are just only if each individual is 'entitled' to their goods. According to such a theory of 'justice in holdings', people are 'entitled to [their holdings] by the principles of justice in acquisition and transfer, or by the principle of rectification of injustice' (Nozick, 1974, p 153). The acquisition and transfer of an individual's holdings is the foundation of libertarian distributive justice. Property is justly acquired, as Locke argued, by cultivating resources through one's own labour which affords one a private right to that property. It may be justly transferred among contractors through voluntary market exchange. Distribution of goods is therefore just only when an individual's entitlements are voluntarily exchanged to another individual without being coerced by any other entity.

The exacting standards of just distribution unique to libertarianism are controversial. Any 'forced' wealth redistribution is understood to be coercive (Machan, 2001). Taxation, for example, is viewed by libertarians as a violation of individual liberty rights (Gaus, 2000). For, taxation constitutes an unjust transfer of property, and for Nozick (1974), it equates to forced labour. This goes against liberal-egalitarian redistributive theories of primary goods, which see taxation as a means of moving towards equality, in that it secures basic primary goods for the poor and for contributing to public goods shared by all of society, such as public infrastructure. For many, then, a libertarian theory of distribution enables social and economic inequality in that it secures wealth in fewer hands.

Mainstream libertarians, however, would reject the idea that wealth can never be redistributed (Wendt, 2019). Redistribution is defensible from a libertarian perspective only when the means of redistribution, voluntary transactions, are non-coercive. As such, the procedure of justice within libertarianism does not incorporate into its evaluation of justice the inequalities among individuals in a society. Rather, inequalities are understood to be naturally occurring and not the responsibility of other individuals to ameliorate. As Hayek (1998 [1976], p 64) argues, iniquitous outcomes from voluntary market exchanges are justified because they are not intended within the voluntary transactions of market exchange. And to 'demand justice' for these inequities by taxing individuals would be unjust. While individuals are free to give away their wealth to those with less as they see fit, they must not be coerced to do so by any person or entity. Tomasi, for example, reminds us that many left-leaning libertarians are committed to a distributive condition in which societies put their resources to the benefit of the least well off (2012).

The libertarian perspective holds that individuals are not means to be used for others' purposes, however, understanding that there is some role for redistribution under strict conditions. But redistribution has nothing to do with justice. Rather, the unequal outcomes of distribution within a society are inconsequential for a theory of libertarian justice. Libertarian 'distributive' justice is in this sense historically constrained in ways that are not seen in other justice theories. The very nature of these constraints means that it is a procedural interpretation of justice that dominates for libertarians. This also means that libertarian approaches to justice offer a very narrow range of remedies for injustice because of how it adheres to strict procedural mechanisms while avoiding material redistribution.

Debates of the tradition

This section traces the historical development of libertarianism to delve more deeply into the conceptions undergirding such a narrow theory of justice. It begins by tracing the roots of mainstream libertarianism, focusing on how self-ownership has been so attached to negative liberty. Given the many criticisms of mainstream libertarianism, the section also presents the diverging positions of left-leaning libertarians that counter mainstream notions of appropriation. To understand the direction of travel in the scholarship, we focus on placing the arguments already outlined into a historical context.

Unlike other political ideologies, the narrow basis for justice within libertarianism has led to a relatively underdeveloped theory. This is so as the systematic pursuit of reducing inequalities is not a stated objective of libertarianism. Discourse on what left-leaning libertarianism has to offer mainstream libertarian theories of justice therefore has not been subjected to ongoing theoretical debate and reflection. By examining key concepts historically within libertarianism in what follows, we tease out the commonalities within libertarianism broadly to show where a libertarian theory of justice may be developed which promotes a more socioeconomic sense of equality. We trace three fundamental concepts in particular, that of self-ownership, property and appropriation, to show the difference in thinking among mainstream and left-libertarians.

Self-ownership is an essential concept within right-libertarianism as that is where its theoretical origins largely derive. To the extent that right-libertarianism holds a moral view, the idea that individuals 'own' themselves is an ethical pillar. The notion that individuals own themselves is often traced to Locke's assertion that individuals have properties in themselves. As Locke (2016 [1689], ch 5, sect 27) famously stated, 'every man has a property in his own person: this no body has any right to but himself. The labour of his body, and the work of his hands, we may say, are properly his'. According to Freeman (2018, p 77), Locke meant not that persons are property themselves, but rather that a property in oneself means that 'no one is born politically subject to another but that each has upon reaching maturity rights of self-rule'. From this view, Locke provides the essential link to liberty of persons germane to both liberalism and libertarianism. But mainstream libertarians take Locke's claim to self-ownership beyond a right to self-rule. Embodying the mainstream libertarian claim that individuals own themselves is Nozick's argument that individuals are akin to possessions themselves. A 'full' right to self-ownership for Nozick is one where each person has 'a right to decide what would become of himself and what he would do, and as having a right to reap the benefits of what he did' (Nozick, as cited in Otsuka, 2003, p 12). When taken as an absolute right, Cohen states that the right of self-ownership is the 'fullest right a person (logically) can have over herself provided that each other person also has just such a right' (Cohen, 1995, p 213). What follows from an absolute property in our person, states Freeman (2018, p 76), is that all individuals 'have absolute power over what we own or acquire consistent with others' ownership rights'. It is this extension of ownership that enables a libertarian right to property.

In that libertarianism sees individual liberty as fundamental, the individual right to property is extended from this by connecting self-ownership rights with rights of private property ownership. For mainstream libertarians, ownership rights must be protected as the most important right of liberty. For, it is through the possession of property that an individual realises their liberty, by being able to use their possessions for their self-preservation (Locke, 2016 [1689]), as well as to transfer and exchange their holdings for individual benefit. As Narveson (1989, p 71) argues, when we understand 'liberty as property', then the libertarian thesis is 'really the thesis that a right to our persons as our property is the sole fundamental right there is'. Thus, Narveson argues that the libertarian right to property is no better epitomised than in the words of Murray Rothbard, who argued that:

> In the profoundest sense there are no rights but property rights. The only human rights, in short, are property rights. ... Each individual, as a natural fact, is the owner of himself, the ruler of his own person. The 'human' rights of the person ... are, in effect, each man's property right in his own being, and from this property right stems his right to the material goods that he has produced. (Rothbard, 1981, p 238)

For mainstream libertarians, the right to property is the right around which all political and philosophical answers revolve. As mentioned, justice is realised when the process of just acquisition and transfer in holdings is observed; a process in which all individuals are not coerced into a mutual exchange of their property or wealth. That liberty is defined by not interfering with an individual's right to property, however, leaves a complicated if not implausible condition for justice from a mainstream libertarian perspective. Particularly the libertarian view regarding how property has come to be justly acquired historically has not been satisfactorily explained.

Mainstream libertarians contend that holdings are just when the acquisition and transfer of those holdings are legitimate. That is, when an individual is not coerced into transferring their property by someone or thing and when it was legitimately acquired through possession. But how do individuals legitimately acquire property in the first instance which can then later be justly transferred? Many go back to Locke, who argued that a resource becomes an individual's property when they invest their labour into that resource. The right of the individual to possess that resource becomes a legitimate right to property, a legitimate right to own property in something. This process is just, Locke argued, only when the individual appropriating a given resource does so by cultivating only what they are able to use and no more, so as to leave resources for others. This so-called Lockean proviso justifies an 'original acquisition' that, while largely contested today, has also been central to libertarian notions of just acquisition and transfer.

While Nozick himself notes that Locke's labour theory – which sees the mixing of one's labour into a resource as the means of coming to own a resource – does not necessarily constitute a right of appropriation, Nozick's 'entitlement' theory nonetheless relies upon the premise of historical acquisition in property holdings as the justification for rights of private property. Nozick adjusts the Lockean proviso to instead 'require that no one can be made worse off as a result of use or appropriation, compared with a baseline of non-use or non-appropriation' of Locke (van der Vossen, 2019, np). Nozick would suggest, like Locke, that the appropriation of property ought to be based only on what one requires. However, if it is found out that other individuals became worse off from an individual's acquisition, this constitutes an injustice to others' rights to equally acquire property. The answer for Nozick is that some rectificatory action must be taken to compensate those who are affected by an unjust transaction.

Without further delving into debates over just acquisition, it is worth noting that many within, much less outside of, libertarianism see the historical acquisition of property as, at best, a grey area that remains to be adequately explained. Simply put, the mainstream libertarian view on justice is one maintaining that rights of liberty, property being the primary liberty, are legitimately held and cannot be redistributed by anyone other than the owner. Libertarianism, more so than liberalism, holds that justice is rooted within the protection of negative rights of liberty. By maintaining that rights of property are historically grounded, mainstream libertarianism obscures the consequences of acquisitions on other individuals' freedom, independence and ultimately equality, by prioritising only the individual right of contract to property (Freeman, 2018). The naturalised process of self-ownership means that some will over time develop different abilities, motivations or inherit different primary goods. Rather than viewing this as something to correct, the libertarian justice framework accepts the natural distribution of these goods over time. The procedure of just distribution within libertarianism, therefore, separates itself from theories of justice which seek to mitigate inequalities in society.

Mainstream libertarianism as understood so absolutely would surely constrain any theory of justice rooted in social equality. But libertarianism is not contained within right-leaning, mainstream interpretations. There exists a range of libertarian perspectives, which to varying extents accept some or almost none of the positions held by right-libertarianism. At the absolute core of any libertarian perspective is that individual liberty is inviolable. Identifying *how* individual liberty is best preserved is where libertarian perspectives diverge. While individual liberty is a priority for right and left libertarians, the range of perspectives on appropriation and ownership differ markedly. What follows is a brief summary of the ways in which libertarian positions differ from the mainstream concerning these concepts.

One can see the range of libertarian perspectives when looking at the differing interpretations of the process of appropriation. The logic of appropriation, the right to cultivate a natural resource thus making it into one's property, is strongest for right-wing libertarians, a position that moves towards outright rejection from left-wing libertarians. What can and should be owned is the point of dispute here. Thus, what level of constraints are in place on appropriation is what matters. Somewhat of a middle ground between far right and left positions on appropriation, Wendt (2018) advocates for a 'sufficiency proviso'. According to Wendt, humans are 'project pursuers' who require some level of property to fulfil their pursuits. The sufficiency proviso underscores a 'practice of private property that ... should be designed in a way that makes sure that everyone has sufficient resources to live as a project pursuer' (Wendt, 2018, p 172). On the face of it, a sufficiency proviso requires that everyone must have enough to live as a 'project pursuer', going against a Nozickean interpretation of the Lockean proviso which leaves individuals without property when others appropriate it first. Similarly to Nozick, however, Wendt suggests that the sufficiency proviso is not a positive right in that it does not require a welfare state. Those who gain unjustly must compensate those who were negatively affected by an unjust acquisition of property, an idea in line with Nozick's thinking.

Wendt's sufficiency proviso can then be compared with a further left 'egalitarian' proviso. Steiner's equal shares proviso, for example, argues that all 'persons have a claim right that others do not appropriate more than an equal share of external resources' (1994, p 235). Otsuka (2003) and Vallentyne (2007) argue similarly that all individuals ought to have a claim right 'that others do not appropriate more than is compatible with equality of opportunity' (cited in Wendt, 2018, p 175). Wendt notes, however, that these more egalitarian provisos conflict with the libertarian position that goes against imposing 'harsh restrictions on legitimate project pursuit' (Wendt, 2018, p 176). There is far more debate about what constitutes the appropriate libertarian proviso which cannot be addressed here, but there are a few points worth noting. A left-libertarian position acknowledges that the standard Nozickean position is not sufficient to meet the needs of individuals. This suggests that left-libertarians do not accept that appropriation rights are absolute nor that they are equitable. Rather, left-libertarian critique argues there must be negotiation over how those with few goods are negatively affected by acquisitions and transfers more so than the right-libertarian position affords through the idea of rectificatory compensation. Given that individuals' rights are preeminent for all libertarians, however, it remains difficult to discern a distributive theory of justice which does not violate at least some individuals' property rights when understood as a liberty right.

As the perspectives on appropriation differ, so too do the justifications for the right to property differ along the spectrum of libertarian thinking. Outside of left-libertarian theories of justice specifically, exists a deep history of resistance to the idea of an absolute right to private property. Libertarian socialists, for example, reject most private ownership, instead insisting that personal property respects the liberty of individuals while avoiding the domination of capitalist property relations which are antagonistic to freedom

(Long, 1998, p 305). The abolition of private ownership over the means of production, a left-libertarian position (Chomsky, 2013), is also related to the abolition of the political authority vested in a traditional state. Different left-libertarian positions accept different levels of relationship with the state. Anarchists reject any state intervention, while strands of socialism accept limited roles for the state. The shared libertarian root of anti-authority stems from a disdain for an unauthorised power of the state to redistribute any personal goods from individuals, a violation of personal liberty. There is of course enormous variation on what state intervention is acceptable within libertarianism making it much too difficult to summarise here.

As far as how left-libertarian thinking on property and the state contribute to thinking on justice, the influences are many. Libertarians see a need for individual control over one's person. That notion that all individuals remain free recognises, if only rhetorically, that all individuals are equal. The protection of each individual's liberty is the means by which libertarianism understands equality. Equality is not to infringe on any individual's rights. While left-libertarian thinking does not offer a substantive theory of justice that right-libertarianism does, left-libertarian thinking remains useful as a foundation for critiquing the clear inequities of the mainstream libertarian theory of justice.

We conclude this section with the key weaknesses raised in debates regarding the cognitive reach and potential of left-libertarian thinking. It lacks voice. Libertarian (or other) scholars are silent on promoting its unique characteristics or engaging in contemporary applied critiques of real-world examples. This first weakness is existential. Justice is an intellectual frontline that most libertarian scholars want to avoid, as to do so would be to acknowledge its importance. A refusal to consider intervention as a means limits its appeal. While raising interesting arguments on individual rights and property, left-libertarian thinking in itself avoids proactivity to such an extent that it becomes ineffectual. Urgings remain too often in the abstract. The absence of a unified theoretical framework results in libertarian justice being questioned on its permissive and productive conditions, purpose and existence. Further scrutiny, and ultimately advocacy, is necessary for its development.

Conclusion

In prioritising equality, mainstream libertarianism seems to offer little by way of equity for a theory of justice. While libertarian arguments related to justice – especially related to the rights of the individual and property – can be imagined as commensurable to, for example, Indigenous concerns with community and the collective, the libertarian reduction of justice to only a question of individual liberty means that much that is central to mainstream libertarianism would have to be jettisoned in conceptualising just distributions in the context of postcolonial justice or non-human subjects of justice.

By way of critique, however, left-leaning libertarian principles offer needed counter-perspectives against the rigid conception of justice emanating from mainstream libertarianism. To the extent that mainstream libertarian distributive justice shapes the procedures of justice within liberal capitalism today – the protection of individual property over that of equality – left-leaning libertarian perspectives identify what cannot be justifiable within social, economic, political and environmental relations. Following the previously mentioned debates over protection of property and the role of the state, left-libertarianism requires that we rethink how individual liberty be best protected by reconfiguring our relationships with property and of the state.

Like other 'radical' traditions critiquing mainstream theories of justice, left-libertarianism offers a few key ideas to consider for what more just relations could look like in the distribution of property, access to procedural justice or recognition justice through the lens of libertarianism:

- A critical assessment of how some property holdings have been unfairly acquired historically and how the protection of unjustly acquired property creates social, political, economic and ecological inequalities.
- A sensitivity to how states arbitrate between rights of property and personal rights of individuals and how the effects of such juridical decisions enable greater social and material inequities for particular groups who have been historically marginalised.
- An awareness of how the natural limits to ecological resources present the notion of an absolute right to property in perpetuity untenable.
- Above all, a left-leaning critique of mainstream libertarianism suggests that there is more than one way to advance a more just set of relationships while protecting individual liberties. A thorough rethinking of how individuals' rights are not only protected but understood relationally is fundamental to advancing any theory of justice that seeks to move beyond the limitations of mainstream libertarianism.

There is significant diversity within libertarian thinking resulting from key points of convergence and divergence among libertarian scholars and how this forms a libertarian position on justice. The libertarian approach to justice provides much for reflection in regards to tolerance, respect and individual rights. We would urge justice thinkers to consider the implications of the varying positions on this spectrum when thinking through libertarian distributive, procedural and recognitional justice rooted in rights of property and self-ownership.

References

Berlin, I. (2002) Two concepts of liberty. In H. Hardy (ed), *Liberty*. Oxford, UK: Oxford University Press, pp 167–218.

Bobbio, N. (2005) *Liberalism and Democracy*. London: Verso.

Brennan, J. (2012) *Libertarianism: What Everyone Needs to Know*. Abingdon, UK: Oxford University Press.

Brennan, J., van der Vossen, B. and Schmidtz, D. (eds) (2018) *The Routledge Handbook of Libertarianism*. New York: Routledge.

Chomsky, N. (2013) *On Anarchism*. New York: The New Press.

Cohen, G.A. (1995) *Self-Ownership, Freedom, and Equality*. Cambridge, UK: Cambridge University Press.

Freeman, S. (2018) *Liberalism and Distributive Justice*. Oxford, UK: Oxford University Press.

Gaus, G.F. (2000) Review essay/A libertarian alternative to liberal justice. *Criminal Justice Ethics*, 19(2), 32–43.

Hayek, F. (1998 [1976]) *Law, Legislation, and Liberty: A New Statement of the Liberal Principles of Justice and Political Economy*. New York: Routledge.

Hobbes, T. (1904) *Leviathan: Or, The Matter, Form & Power of a Commonwealth, Ecclesiastical and Civil*. London: C.J. Clay and Sons.

James, S. (2003) Rights as enforceable claims. *Proceedings of the Aristotelian Society*, 103(2), 133–147.

Kaufmann, M., Leroy, P. and Priest, S.J. (2018) The undebated issue of justice: Silent discourses in Dutch flood risk management. *Regional Environmental Change*, 18(2), 325–337.

Locke, J. (2016 [1689]) *Two Treatises of Government*, L. Ward. (ed). Indianapolis, IN: Hackett Publishing.

Long, R. (1998) Toward a libertarian theory of class. In E. Paul, F. Miller and J. Paul (eds), *Problems of Market Liberalism*. Cambridge, UK: Cambridge University Press, pp 303–349.

Machan, T.R. (2001) Libertarian justice. In J. Sterba (ed), *Social and Political Philosophy: Contemporary Perspectives*. New York: Routledge, pp 93–114.

Narveson, J. (1989) *The Libertarian Idea*. Philadelphia, PA: Temple University Press.

Nozick, R. (1974) *Anarchy, State, and Utopia*. New York: Basic Books.

Otsuka, M. (2003) *Libertarianism with Inequality*. Oxford, UK: Oxford University Press.

Rothbard, M. (1981) *Power and Market*. New York: New York University Press.

Steiner, H. (1994) *An Essay on Rights*. Oxford, UK: Blackwell.

Tomasi, T. (2012) *Free Market Fairness*. Princeton, NJ: Princeton University Press.

van der Vossen, B. (2019) Libertarianism. In E.N. Zalta (ed), *The Stanford Encyclopedia of Philosophy*. https://plato.stanford.edu/archives/spr2019/entries/libertarianism/

Vallentyne, P. (2007) Libertarianism and the state. *Social Philosophy and Policy*, 24(1), 187–205.

Waldron, J. (1993) *Liberal Rights: Collected Papers 1981–1991*. Cambridge, UK: Cambridge University Press.

Wendt, F. (2018) The sufficiency proviso. In J. Brennan, B. van der Vossen and D. Schmidtz (eds), *The Routledge Handbook of Libertarianism*. New York: Routledge, pp 169–183.

Wendt, F. (2019) Three types of sufficientarian libertarianism. *Res Publica*, 25(3), 301–318.

3

Cosmopolitan Theories of Justice

Tracey Skillington

Introduction

The ideas traditionally associated with a cosmopolitan model of justice have a long and varied history, stretching from the writings of the early Stoics and Cynics to Immanuel Kant's reflections on revolution, moral universalism and hospitality, to philosophical contemplations on the importance of nurturing a Europe of *peoples* committed to a 'good peace' in the post-Second World War period (for example, Arendt, 1950) or more recently, commemorations of United Nations Day (24 October) reaffirming the cosmopolitan purpose of the UN Charter. In more contemporary phases, this history has taken Kant's critical reworkings of the ideas of the Cynics and Stoics into new areas of investigation, including the ethics of migration (Heath and Cole, 2011), the rights of refugees (Huber, 2017), war and peace (Fabre, 2016) and global governance (Moellendorf, 2005). Among those ideas with an enduring presence in the 'cosmopolitan imagination' (Delanty, 2005) are the universal moral affiliations of 'the citizen of the world' (*kosmo politês*) where class, status and local origin are seen as secondary markers of identity.[1] The first form of affiliation for the cosmopolitan citizen is affiliation with the community of humanity. For the Stoics, each of us dwells in two communities simultaneously: the community of our residence (territorially grounded) and the community of human argument and aspiration (trans-territorial). The latter, in Seneca's words, is a type of community 'in which we look neither to this corner nor to that, but measure the boundaries of our nation by the sun'.[2] Hierocles, a Stoic of the 1st and 2nd centuries AD, saw our affiliations with the community of humanity as supported by a series of concentric circles of belonging that stretch far beyond the local. The most outer circle of this affiliation being that of humanity as a whole. The primary moral task of the citizen of the world, according to Cicero, is to always draw this circle 'towards the center' of our shared existence.[3]

Undoubtedly, the moral affiliations of the Stoics and Cynics shaped Kant's vision of the cosmopolitan in the 18th century and would prove to have a lasting influence on struggles against absolutism, debates on the ethics of European colonialism, the Reformation, as well as other geopolitical, cultural and social realities of the day (Reich, 1939). In his assessment of the moral and practical foundations of our 'communal possession of the earth's surface', for instance, Kant (1983 [1795]) pointed to the connectedness of the world's populations as one community, where 'a violation of rights in one part of the world is felt everywhere'.[4] In these circumstances, some degree of interaction between strangers is unavoidable and raises several moral issues regarding the reception of those who have fallen victim to the 'inhospitable actions of the civilized and especially of the commercial states', including issues of responsibility and duties of care. In situations of growing planetary interdependency, developing a civil society with the capacity to 'administer justice universally' was considered by Kant (2010 [1797]) to be essential. For inspiration on this front, Kant looked to the writings of such thinkers as Abbe de Saint-Pierre (1713) and Jean-Jacques Rousseau (1954 [1762]). Both writers stressed the importance of a new legal and political order of justice with the capacity to transform lawless relations between peoples and states into civil relations of perpetual peace. The influence of Bentham's (1939 [1786–1789]) ideas of an international court to resolve conflicts between peoples is also evident in the writings of Kant. More than any of these thinkers, however, it was Kant who advocated for the rights of world citizens, seeing those rights as providing the strongest justification for the development of a new global order of justice.

Enacting 'a history with a cosmopolitan purpose', where universal rights, perpetual peace, inclusivity and openness to difference bind the actions of all members of the democratic community, remains the structuring normative intuition of the cosmopolitan justice tradition even today. The value of this history, however, is widely debated in light of persisting problems of inequality, racism, gender violence, war, populism, natural resource destruction and unmitigated climate change, all of which bear heavily on modernity's cosmopolitan aims (perpetual peace, universal rights) and commitments to a global democratic order. The following discussion accounts for a number of different types of cosmopolitanism, the relevance of which bear on many of the issues raised here. The general tenets of each will be explored first before their applicability to a variety of current justice concerns is considered.

Varieties of cosmopolitanism

Legal cosmopolitanism

Legal cosmopolitanism is concerned with the legal status of individuals as human beings, rather than as citizens of specific states. In Kant's political theory, cosmopolitan law is the third category of public law (constitutional law is the first and international law, the second). This third level of law proposed by Kant is deemed necessary to protect the rights and dignity of all individuals. It advocates for legal protections and obligations beyond a traditional state-centric model of international law. Kant drew attention to the limitations of a law between states or even that between states and their citizens, proposing as a supplement, a cosmopolitan law between 'citizens of the earth' or a universal state of humankind. For Kant (1983 [1795]: 357), a chief concern when developing the idea of cosmopolitan law is that it embodies 'the conditions of universal hospitality'.

For Kant, hospitality is 'the right of a stranger not to be treated with hostility because his arrival on someone else's soil' (Kant, 1983 [1795], p 358). However, hospitality is understood as a right to visit. No one has a right to settle on the soil of another community since this right can only be established through a treaty agreement among all concerned parties. Kleingeld (1998, p 83) believes this peculiar reading of hospitality on the part of Kant was in the interests of supporting a limitation on the rights of colonialist aggressors and moves to take possession of land overseas without any regard for the claims of native populations. Kant's hospitality, Kleingeld (1998, p 76) argues, is aimed at supporting the sovereignty of Indigenous peoples against unwanted European encroachment and, therefore, could be seen as equally relevant today as a legitimate defence against large-scale global corporate mining, deforestation, gas and oil exploration projects that detrimentally affect the lives and livelihoods of settled communities (human and non-human alike). Some, however, disagree with Kleingeld's interpretation of Kant (Mignolo, 2009). (For further discussion, see the section 'The contemporary analytic relevance of cosmopolitanism'.)

Political cosmopolitanism

'Cosmo-politics' is a politics of co-habitation with worldly companions. It is popularly conceived in the literature as 'an alternative to nationalism' (for example, Cheah and Robbins, 1998) although there are exceptions. Beck and Levy (2013), for instance, document the growing cosmopolitan affiliations of the nation state in response to global economic, political and legal changes. They point to increasing challenges to traditional interpretations of the state as a natural repository of political legitimacy. For example, the influence of state-transcending human rights norms and international treaties with the potential to circumscribe the power of the state. In this setting, states renegotiate the social, political and legal basis of their legitimacy in terms of a willingness to embrace cosmopolitan norms and recognise the rights of individual members, as well as the collective rights of distinct communities (for example, Indigenous peoples) to self-determination. Respect for persons and peoples, as well as institutional mechanisms that formally protect their freedoms, come to be prioritised. Indeed, respect is also an essential part of the cosmopolitan framework of a wider, integrated Europe (legally inscribed), as much as it is a component of the shared political ethos of its citizens (that is, definitions of European citizenship). In *Bounds of Justice*, O'Neill (2000) explains this form of cosmopolitanism as one that argues in favour of a universal justice that takes seriously the role of multiple political institutions in its realisation. Similarly, David Held (1995) advocates a cosmopolitan democracy that brings 'overlapping communities of fate' together to address major justice issues of common concern, including those arising in relation to climate change, rising temperatures, the breakdown of food chains, escalating pollution and loss of biodiversity. The following principles typically inform a cosmopolitan model to democratic governance:

1. The equal moral worth and dignity of all persons.
2. Active rights agency.
3. Personal responsibility and accountability.
4. Democratic reason.
5. Respect for difference.
6. Free and equal consent.
7. Collective decision-making through direct or representative democratic procedures.
8. Avoidance of serious harm.

While many components of political cosmopolitan thinking overlap with those of other traditions, most notably, liberalism, there are also important differences between them. For instance, cosmopolitanism shares with liberalism a concern for the right to life, liberty and equality, as well as principles of autonomy, respect, civic solidarity, democratic constitutionalism and the free consent of the governed. A classic liberal mantra is that of Locke's 'life, liberty and property', which, from a cosmopolitan perspective, is somewhat problematic on account of its interpretation of 'entitlement' as deriving chiefly from a human investment in material things (for example, land and other resources essential to life) and the idea of liberty as tied to property (as a private holding) rather than any unconditional sources of entitlement (that is, universal right). That said, it is difficult to identify 'pure' contemporary applications of cosmopolitan ideas or texts where authors do not also draw on state liberal, constitutional or communitarian approaches (for example, Erskine, 2008). Most embody what Appiah (2019) describes as a cosmopolitanism that is focused on 'both the near and far' dimensions of social and political worlds.

In terms of its practical application, a great deal of the literature on political cosmopolitanism over the last three decades has focused on the following themes: the borders of contemporary Europe, the reconfiguration of sovereignty (particularly in the post–Second World War period), the status of the refugee, migration and climate change.[5] All of this literature refers, directly or indirectly, to the work of Kant. In particular, Kant's reflections on the rights of the stranger to hospitality or to be recognised as a participant in a world republic. The right to have rights is one recognised to some degree by the democratic sovereign state but, as cosmopolitans argue, an additional layer of institutional recognition is required beyond the state to protect the unconditional right of every human being to belong somewhere (that is, the earth's surface as a 'possession in common').[6] History informs us that we cannot always trust states to fulfil these obligations and treat the individual with respect.[7] To overcome a heavy dependency on the goodwill of states in this regard, cosmopolitans propose locating the legal grounds of rights partially in a different order of belonging, such as a transnational legal and political order of democratic rights (Habermas, 2015, p 60).

While critical of some aspects of Kant's interpretation of cosmopolitanism, Habermas (1997, p 135) does admit that 'the moral universalism that guided Kant's proposals [on cosmopolitanism] remains the structuring normative intuition' of transnational democracy even today. Similarly, Apel (1997, p 100) expresses a desire to retain Kant's primary concern with universalism but does draw attention to certain contradictory elements in Kant's philosophy of history. For example, the claim that the bringing about of a cosmopolitan order is a moral duty and, simultaneously, the guaranteed outcome of a natural mechanism. As Apel points out, a cosmopolitan order has to be consciously brought into existence. Its emergence is, therefore, not self-evident or inevitable. In *Conflict of the Faculties*, Kant (1979, p 84) raises the French Revolution to the level of a 'historical sign' of humanity's potential for moral progress. As a founding act of moral accomplishment, this revolution, Kant argues, is likely to sustain commitments to a rational constitutional democracy into the future. However, as Apel (1997) and indeed Habermas (2001, p 770) observe, the cohesive functions of such a democracy also depend crucially on a communicatively achieved constitution-making, as well as processes of democratisation that are ongoing rather than the product of once-off revolutionary events. If the relevance of Kant's universal concerns is to be preserved then the conditions for the realisation of a cosmopolitan order of justice must not be seen as deriving from singular or even essentially European histories but, rather, from ongoing processes of open communication and collective societal learning. That is, one where unhindered debate among all affected communities (and differing worldviews) remains constant. Once these clarifications are established, the ongoing relevance of Kant's vision of the cosmopolitan to the cognitive and normative orders of modernity is more evident.

Moral cosmopolitanism

Moral cosmopolitanism argues that the same moral standards ought to apply to all individuals regardless of where they reside. The justice of this position is defined as equality across borders. Scholars such as Martha Nussbaum (1996), for instance, call for a cosmopolitan moral education that highlights the constructed nature of borders between peoples. However, this plea, with its emphasis on what is common to all, has been subject to critique on the grounds that it could potentially be used to downplay morally relevant differences (for example, Mignolo, 2009) and reassert western bias in formulations of 'our common humanity'. Sameness, critics argue, may be construed as a license to extend recognition chiefly to those who resemble the European ideal. Distinct otherness may pose a problem in this instance. On the other side are those who claim proposals to treat all humans as equal on the basis of a shared capacity for reason are not designed to discriminate (on the grounds of race, class, gender, ethnicity, nationality, sexuality, and so on). Kant attempts to answer issues raised by the 'practical experience' of a history 'pathologically compelled' by wars, plunder, exploitation, discrimination and inequality by differentiating it from the possibility of one 'with a cosmopolitan purpose' (1963 [1784]). Even so, there does appear to be some contradictory elements at the core of Kant's notion of the cosmopolitan. On the one hand, there is a reference to the vast expanse of a global world order, where the reach of nature and the 'cosmos' is boundless. On the other hand, there is a reference to the all-too-human idea of a bounded or circumscribed polis where humans are said to be elevated 'above nature' through the civilising process. The polis is thereby created as an artificial entity designed to wield tendencies towards war and impose peace. Peace thus comes to be seen as the contingent element of a democratic contractual arrangement within which we acknowledge each other as equals but only under these conditions and the watchful eye of the demos.

Kant did distinguish pure moral cosmopolitanism (transcendentally free) from the empirical subject (and the messiness and particularity of social interpretation). In other words, the interpretation and social application of principles of cosmopolitanism, such as those explored here, continue to evolve via conflict-led learning, or what Adorno refers to as engagements with the 'non-identical'. Following Kant, Hale and Held (2011), Strydom (2012) and Beck and Sznaider (2006) all draw attention to the importance of the dialectic established between cosmopolitanism as a cognitive and normative ideal and real or embedded processes of 'cosmopolitanisation', where its principles take on a social persona and guide the direction of institutional learning (economic, social, legal and political). It should be noted that for Beck, in particular, the latter is rarely a voluntary endeavour but one more often provoked by situations of deepening crisis and the sheer necessity of cooperation in the interests of achieving collectively beneficial outcomes. In this dimension of his thinking on cosmopolitanism, Beck does depart somewhat from Kant who sees this process more as a natural progression.

Common to all, however, is the notion that cosmopolitanism cannot be rigidly standardised as contradictory or blind to the non-western other if its relevance continues to be explored 'dialogically' in new contexts of discovery (or through what Benhabib [2007] refers to as 'democratic iterations'). Similarly, moral cosmopolitanism when considered in terms of the empirical finitude of this planet's resources, encourages us to inhabit its territories in a manner that accommodates the needs of others. Kant highlights how the physical facts of the Earth's geography compel us to be cosmopolitan in our orientation to and co-existence with the other. All cosmopolitan theorists embrace this distinct worldview, sometimes referred to as 'world openness'. To act upon and know this world from a cosmopolitan perspective, therefore, is to continuously question how the conditions and duties of our co-existence can be improved.

Cultural cosmopolitanism

Cultural cosmopolitanism celebrates cultural variety both within and across world communities. As the congruence between ethnicity, nationality, citizenship and territory continues to decline, traditions of cultural cosmopolitanism flourish where the identity of the individual is forged more on the basis of trans-territorially available cultural resources rather than those inherited exclusively from a specific place or past. Cultural cosmopolitanism recognises the rights of the individual to exercise freedom of cultural expression (Appiah, 1997) and draw a sense of belonging to communities that extend far beyond the local. In that, it implicitly embraces principles of universal justice (for example, autonomy, tolerance, respect for difference).

To explain these components of cultural cosmopolitanism, Appiah (1997, p 618) quotes Stein who in 1936 famously described how 'America is my country and Paris is my hometown'. For Delanty (2005, pp 405–419), cultural cosmopolitan affinities, such as those expressed by Stein, are central to the emergence of various post-national solidarities and a growing allegiance to transnational justice movements (for example, Climate Justice Alliances, Black Lives Matter, Me Too, and so on). Similarly, Beck et al (2016, pp 111–128) explore the significance of cosmopolitan memory projects (for example, the Holocaust or Rwanda genocide) as an expression of cultural cosmopolitanism and support for the further advancement of post-war expressions of legal, political and cultural solidarities (for example, UN Outreach programmes on the Holocaust and Rwanda genocide, the International Criminal Court). Symbolic forms of representation of justice may vary across different varieties of cosmopolitanism, as they are noted here. However, what remains central is a focus on the *kosmo polités* (that is, citizen of the world).

Cosmopolitan approaches to distributive and procedural justice

Distributive justice

The standard assumption in cosmopolitan thinking is that the 'scope of justice' specifying *to whom* goods should be distributed has a global dimension (inclusive of members of all communities). For cosmopolitans, duties of distributive justice should not be constrained or limited by state borders. Rather, the appropriate subject matter of justice is global. If we accept the claim that all individuals have moral value and that this moral worth, equally shared by all members of humanity, generates moral duties that are binding on everyone, it would be contradictory to assume that these duties apply only in interactions with fellow citizens or co-nationals (Caney, 2005). Universalist considerations imply that the scope of justice must be considered in broader terms, irrespective of ethnicity, gender, class, nationality, age or the individual's 'moral personality' (Rawls, 1999, pp 11, 17, 442–446). Underlining this argument is a cosmopolitan understanding of justice as globally applicable.

More generally, one may distinguish 'radical' from 'mild' varieties of cosmopolitan distributive justice. In the case of radical varieties, global principles of justice are prioritised over state ones. Beitz (1999, p 182), for instance, explains how 'state boundaries can have derivative' relevance on issues of distributive justice but 'they cannot have fundamental moral importance'. Similarly, Caney (2001) argues that equality of opportunity cannot be determined by state membership or geographical location when it comes to questions of distributive justice. For equality of opportunity to thrive, rights protections must be secured for all. One of the most profound violations of global cosmopolitan distributive justice is unmitigated climate destruction (Moellendorf, 2009). Climate change violates practically every egalitarian principle of human rights law and, therefore, requires redress through a series of concrete measures. One such proposal entails a dual application of cosmopolitan distributive justice principles. First, at a transnational level where a confederation of high polluting states acknowledges its historical responsibilities for inflicting hardships on others, as well as the ongoing capabilities of its members to address these harms and offer assistance to its primary victims. Second, at the global level where a new global resource tax regime could be introduced to accumulate revenues from high polluters around the world and redistribute funds among those most adversely affected by the effects of deteriorating climate conditions. Theorists such as Nussbaum (2006) and Pogge (2008) have explored the merits and potential shortcomings of this model of cosmopolitan distributive justice.

Procedural justice

Cosmopolitan democracy generally supports the idea of procedural equality among all peoples and promotes models of governance designed to address any disconnect that may exist between decision-making arrangements and democratic rights. A significant literature focuses on how this disconnect plays out in relation to various climate change concerns. In principle, a transnationally shared ecological fate ought to lend itself easily to democratic decision-making across borders. However, this proves not to be the case for various reasons, including the dominance of a nation state mentality, an unequal distribution of power, restricted opportunities for publics to confront major sources of harm and initiate far-reaching change (Skilllington, 2017; 2020). As the circumstances of justice continue to evolve and ecological conditions decline further, the need to accommodate a more multi-levelled governance structure increases (Caney, 2005; Held, 2010b; Dryzek and Pickering, 2019). With competition for scarce resources intensifying, a transnational deliberative regime, building positively on structures already established (at the UN or Arctic Council level, for instance), is needed to secure a more effective resource management regime for the future, one based on a recognition of the legitimacy of a broad range of interests. Dryzek (2017) proposes chambers of transnational discourse while Bohman (2010) suggests deliberative intermediaries to maintain a healthy correspondence between the powers of peoples and elected representatives across borders. For Held (2010a), an elected global parliament is also desirable to address concerns that are neither wholly domestic nor international, but the concern of the 'universal community' as a whole. Practical problems, including more intense and frequent flooding, storm surges and rising global temperatures threaten the territorial belonging of multiple communities (within and across state borders) and offer the strongest practical justification for a partial decoupling of decision-making authority from the sovereign bounds of nation states to ensure that the burdens of *globally* induced climate change do not fall disproportionately on some citizens of the world.

The contemporary analytic relevance of cosmopolitanism

Current meta-reflections on the social world move the cosmopolitan paradigm in new directions. For instance, renewed interest in the possibilities of a 'decolonial cosmopolitanism'. In order to understand how a 'Eurocentred cosmopolitanism' continues to discriminate and exclude, one must see it not as an end in itself, Mignolo (2009) argues, but, rather, a first step towards embracing a decolonial cosmopolitanism. For Mendieta (2009), this means revisiting the question of how traditional links established between the cosmos and the European polis were asserted historically in the interests of power and global domination. Mignolo (2009) argues that cosmopolitanism today must take account of these historical legacies as well as two basic characteristics of the contemporary world: the highly constructed nature of ongoing divisions between 'developed' and 'developing/underdeveloped' world regions, and, second, how this situation emerged on the basis of a gross misuse of political and economic power. A decolonial cosmopolitanism is one that embraces cosmopolitan values but in ways not dictated by some grand imperial design or a global order of linear thinking. Rather, it accommodates pluriversal visions of human history and traditions of thinking. A decolonial cosmopolitanism does not strive to be homogeneous, as Mignolo (2009) points out, but embraces the possibilities of a heterogeneous cosmopolitanism on the basis of a learning from global episodes of historical wrongdoing (for example, more recent efforts to repatriate ancient artefacts stolen by former colonialists and now housed in major European museums, see #BlackArtsMatter). One proposal in this regard is to celebrate ancient affinities between the polis and the cosmos among non-European historical civilisations, for instance, the Andean civilisations, the Tiwanaku and Incas Empires, or the Kingdom of Benin in West Africa and, in doing so, honour non-European and non-western traditions of cosmopolitan thought, memories, ways of living, labouring, believing, interacting with nature, and so on.

Van Dooren (2016) develops a decolonial cosmopolitanism in a somewhat different direction, extending Derrida's views on cosmopolitan hospitality to a consideration of relations with non-human others. For Derrida, cosmopolitan hospitality is a constant corrective that pushes us towards new normative horizons, or a new 'cosmopolitan imagination' (Delanty, 2009) of planetary belonging. However, this is a hospitality that does not always arise spontaneously but is regularly prompted by the need to address some immanent crisis, such as war in the Ukraine, or the current critical loss of biodiversity and its relation to the emergence of a range of new deadly risks to health. In its most recent Biodiversity Strategy for 2030, the EU reflects on the need to extend horizons of the common good to include biologically diverse communities, to work with (rather than against) a nature upon which we depend for 'the food we eat' and 'the air we breathe'. A nature, it adds, which is as important for our mental and physical well-being as it is to society's ability to cope with global challenges into the future (European Commission, 2020). By engaging with the implications of more-than-human species entanglements and reassessing how exploitative and harmful relations with non-human nature relate to wider, more historically embedded realities (for example, the emergence of the Anthropocene, the Capitalocene, histories of colonialism, the COVID-19 global pandemic, and so on), government finds itself unable to escape certain decolonial imperatives. It is forced to think again about legacies of destruction that are continuously renewed (that is, the deeper costs of an ongoing colonisation of 'res nullius' territories and their resources, organised violence against mass farmed animals, the effects of melting ice, disappearing food supplies and wildlife populations) and their contribution to the expanding risks of wholesale ecosystem collapse (Skillington, 2020a; 2020b). While unlimited hospitality towards this nature may be an impossible hospitality (Derrida, 2001) as an ideal, it does compel deeper reflection on the question of who can legitimately be excluded from thicker spheres of cosmopolitan justice and why. In that, it triggers further critical engagement with obligations to recognise the normative status of the other as an autonomous agent deserving of recognition, hospitality and various other duties of care. For Derrida (2001, p 11), cosmopolitan justice is a matter of opening ourselves up to hospitalities yet to come, the full meaning of which cannot be entirely anticipated at this point. As hospitality, tolerance, inclusion and respect are enacted most clearly in relational exchanges with the other, including those not yet had, their meaning and significance cannot be wholly determined at this point in time. Equally, those to whom we extend hospitality and respect (including other earthly inhabitants, human and non-human) cannot be settled in advance. Hence, the need to defend the ongoing relevance of a principle of cosmopolitan world openness in newer deliberations on justice (Skillington, 2023).

Conclusion

In conclusion, we may note how dialogical readings of the contemporary relevance of cosmopolitan principles attempt to address the sociological deficits that mark earlier applications (for example, allegations of Eurocentrism, cultural blindness, imperialism, speciesism, racism, colonial exploitation, and so on) and enable a practical overcoming of these limitations. Emerging from such critical dialogues are newer readings of the ongoing relevance of cosmopolitan models of justice in new contexts of reception (for example, the relevance of plural conceptions of justice, tolerance and respect to the ethical and legal status of the non-human subject, non-western histories of atrocity, the actualisation of peace in war-torn Europe, pluriversal worldviews). Through 'democratic iteration' (Benhabib, 2007), cosmopolitan norms come to be mediated with newer critical insights on the enduring nature of problems of Eurocentrism, resource-grabbing, war, racism, violence and exclusion. Cosmopolitan principles are redeployed to consider how old problems persist in new forms (for example, CO_2 colonialism). What remains constant, however, is a focus on the moral, political and legal status of 'the citizen of the world' who in sharing planet Earth in common with others is bound by duties of hospitality and respect for global strangers. When considering the implications of these commitments, the realisation is that the cosmopolitan tradition must engage reflexively with contemporary decolonial imperatives. For instance, the need to decouple understandings of the relationship between the cosmos and polis from strictly European or western interpretive traditions and substitute universal with pluriversal visions of the cosmopolitan. In the process, new constellations of cosmopolitan justice are created in dialogue with the imperatives of a changing world.

Notes

[1]   See *Diogenes Laertius, Life of the Eminent Philosophers*, quoted in Nussbaum (2010, p 29).
[2]   See Long and Sedley (1987, p 431), quoted in Nussbaum (1997, p 6).
[3]   Nussbaum (2010, p 31); see also Griffin and Atkins (1991).
[4]   Quoted in Habermas (1997, pp 113–154).
[5]   On the topic of borders, see Balibar (2009, pp 101–114); Levy and Sznaider (2006, pp 657–676). On the status of the refugee and migration trends, see Benhabib (2004) and Kymlicka (2010, pp 97–112).
[6]   See Kant, DoR 6, 2010 [1797,] p 262.
[7]   For example, when the state, as an instrument of legal and political justice, becomes 'perverted' and obstructs the course of normal procedures of justice, as when a dictatorship emerges.

References

Abbe de Saint-Pierre, C.I.C. (1713) *Projet pour rendre la paix perpétuelle en Europe*, 3 vols. Utrecht, The Netherlands: A. Schouten.

Apel, K.O. (1997) Kant's 'Towards Perpetual Peace' as historical prognosis from the point of view of moral duty. In J. Bohman and M. Lutz-Bachmann (eds), *Perpetual Peace: Essays on Kant's Cosmopolitan Ideal*. Cambridge, MA: The MIT Press, pp 79–110.

Appiah, K.A. (1997) Cosmopolitanism compatriots. *Critical Inquiry*, 23(3), 617–639.

Appiah, K.A. (2019) The importance of elsewhere: In defense of cosmopolitanism. *Foreign Affairs*, March/April. https://www.foreignaffairs.com/articles/2019-02-12/importance-elsewhere

Arendt, A. (1950) Peace or armistice in the near east? *The Review of Politics*, 12(1), 56–82.

Balibar, E. (2009) *We, the People of Europe? Reflections on Transnational Citizenship*. Princeton, NJ: Princeton University Press.

Beck, U. and Levy, D. (2013) Cosmopolitanized nations: Reimagining collectivity in world risk society. *Theory, Culture & Society*, 30(2), 3–31.

Beck, U., Levy, D. and Sznaider, N. (2016) Cosmopolitanisation of memory: The politics of forgiveness and restitution. In M. Nowicka and M. Rovisco (eds), *Cosmopolitanism in Practice*. Oxon, UK: Routledge, pp 111–128.

Beck, U. and Sznaider, N. (2006) Unpacking cosmopolitanism for the social sciences: A research agenda. *The British Journal of Sociology*, 57(1), 1–23.

Beitz, C. (1999) *Political Theory and International Relations*, 2nd edn. Princeton, NJ: Princeton University Press.

Benhabib, S. (2004) *The Rights of Others*. Cambridge, UK: Cambridge University Press.

Benhabib, S. (2007) Democratic exclusions and democratic iterations: Dilemmas of 'just membership' and prospects of cosmopolitan federalism. *European Journal of Political Theory*, 6(4), 445–462.

Bentham, J. (1939 [1786–1789]) *Plan for a Universal and Perpetual Peace 1786–1789*. London: Peace Book Company.

Bohman, J. (2010) *Democracy Across Borders*. Cambridge, MA: The MIT Press.

Caney, S. (2001) Cosmopolitan justice and equalizing opportunities. *Metaphilosophy*, 32(1), 113–134.

Caney, S. (2005) *Justice Beyond Borders: A Global Political Theory*. Oxford, UK: Oxford University Press.

Cheah, P. and Robbins, B. (1998) *Cosmopolitics: Thinking and Feeling Beyond the Nation*. Minneapolis, MN: University of Minneapolis Press.

Delanty, G. (2005) The idea of a cosmopolitan Europe: On the cultural significance of Europeanization. *International Review of Sociology*, 15(3), 405–421.

Delanty, G. (2009) *The Cosmopolitan Imagination: The Renewal of Critical Social Theory*. Cambridge, UK: Cambridge University Press.

Derrida, J. (2001) *On Cosmopolitanism and Forgiveness*. Oxon, UK: Routledge.

Dryzek, J. (2017) The forum, the system and the polity: Three varieties of democratic theory. *Political Theory*, 45(5), 610–636.

Dryzek, J. and Pickering, J. (2019) *The Politics of the Anthropocene*. Oxford, UK: Oxford University Press.

Erskine, T. (2008) *Embedded Cosmopolitanism*. Oxford, UK: Oxford University Press.

European Commission (2020) *Biodiversity Strategy for 2030*. https://ec.europa.eu/info/strategy/priorities-2019-2024/european-green-deal/actions-being-taken-eu/eu-biodiversity-strategy-2030_en

Fabre, C. (2016) *Cosmopolitan Peace*. Oxford, UK: Oxford University Press.

Griffin, M. and Atkins, E. (1991) *Cicero: On Duties*. Cambridge, UK: Cambridge University Press.

Habermas, J. (1997) Kant's idea of perpetual peace, with the benefit of two hundred years' hindsight. In J. Bohman and M. Lutz-Bachmann (eds), *Perpetual Peace: Essays on Kant's Cosmopolitan Ideal*. Cambridge, MA: The MIT Press, pp 113–154.

Habermas, J. (2001) Constitutional democracy: A paradoxical union of contradictory principles. *Political Theory*, 29(6), 766–781.

Habermas, J. (2015) *The Postnational Constellation: Political Essays*. Cambridge, UK: Polity.

Hale, T. and Held, D. (2011) *Handbook of Transnational Governance*. Cambridge: Polity.

Heath, C. and Cole, P. (2011) *Debating the Ethics of Immigration: Is There a Right to Exclude?* Oxford, UK: Oxford University Press.

Held, D. (1995) *Democracy and the Global Order: From the Modern State to Cosmopolitan Governance*. Cambridge, UK: Polity.

Held, D. (2010a) Reframing global governance: Apocalypse soon or reform! In G. Wallace Brown and D. Held (eds), *The Cosmopolitan Reader*. Cambridge, UK: Polity, pp 293–311.

Held, D. (2010b) Principles of cosmopolitan order. In G. Wallace Brown and D. Held (eds), *The Cosmopolitan Reader*. Cambridge, UK: Polity, pp 229–243.

Huber, J. (2017) Cosmopolitanism for Earth dwellers: Kant on the right to be somewhere. *Kantian Review*, 22(1), 1–25.

Kant, I. (1963 [1784]) On history. In *Idea for a Universal History from a Cosmopolitan Point of View*, translated by L. White Beck. London: The Bobbs-Merrill Co.

Kant, I. (1979) *The Conflict of the Faculties*, translated by M.J. Gregor. New York: Abaris.

Kant, I. (1983 [1795]) *Perpetual Peace and Other Essays*, translated by T. Humprey. Indianapolis, IN: Hackett.

Kant, I. (2010 [1797]) *Metaphysics of Morals*. Cambridge, UK: Cambridge University Press.

Kleingeld, P. (1998) Kant's cosmopolitan law: World citizenship for a global order. *Kantian Review*, 2, 72–90.

Kymlicka, W. (2010) The rise and fall of multiculturalism? New debates on inclusion and accommodation in diverse societies. *International Social Science Journal*, 61(199), 97–112.

Levy, D. and Sznaider, N. (2006) Sovereignty transformed: A sociology of human rights. *British Journal of Sociology*, 57(4), 657–676.

Long, A. and Sedley, D. (1987) *The Hellenistic Philosophers*. Cambridge, UK: Cambridge University Press.

Mendieta, E. (2009) From imperial to dialogical cosmopolitanism. *Ethics and Global Politics*, 2(3), 241–258.

Mignolo, W. (2009) Cosmopolitanism and the decolonial option. In T. Strand (ed), *Cosmopolitanism in the Making*. Berlin, Germany: Springer, pp 110–125.

Moellendorf, D. (2005) Persons' interests, states' duties and global governance. In G. Brock and H. Brighouse (eds), *The Political Philosophy of Cosmopolitanism*. Cambridge, UK: Cambridge University Press, pp 148–163.

Moellendorf, D. (2009) Justice and the assignment of the intergenerational costs of climate change. *Journal of Social Philosophy*, 40(2), 204–224.

Nussbaum, M. (1996) Patriotism and cosmopolitanism. In J. Cohen (ed), *For Love of Country?* Boston, MA: Beacon Press, pp 3–17.

Nussbaum, M. (1997) Kant and Stoic cosmopolitanism. *The Journal of Political Philosophy*, 5(1), 1–25.

Nussbaum, M. (2006) *Frontiers of Justice*. Cambridge, MA: Harvard University Press.

Nussbaum, M. (2010) Kant and cosmopolitanism. In G. Wallace Brown and D. Held (eds), *The Cosmopolitan Reader*. Cambridge, UK: Polity, pp 27–44.

O'Neill, O. (2000) *Bounds of Justice*. Cambridge, UK: Cambridge University Press.

Pogge, T. (2008) *World Poverty and Human Rights*, 2nd edn. Cambridge, UK: Polity.

Rawls, J. (1999) *A Theory of Justice*, revised edn. Cambridge, MA: The Belknap Press of Harvard University Press.

Reich, K. (1939) Kant and Greek ethics. *Mind*, 48, 338–354.

Rousseau, J.-J. (1954 [1762]) *The Social Contract*, translated by W. Kendall. Chicago, IL: The Henry Regnery Company.

Skillington, T. (2017) *Climate Justice & Human Rights*. New York: Palgrave.

Skillington, T. (2020a) Natural resource inequities, domination and the rise of youth communicative power: Changing the normative relevance of ecological wrongdoing. *Distinktion: Journal of Social Theory*, 22(1), 23–43. DOI: 10.1080/1600910X.2020.1775669

Skillington, T. (2020b) Democracy, biodiversity and more than human justice relations: Institutional responses to crisis. *Social Sciences*, 9(166), 1–16.

Skillington, T. (2023) Thinking beyond the ecological present: Critical theory on the self-problematisation of society and its transformation. *European Journal of Social Theory*, 26(2), 236–257. https://doi.org/10.1177/13684310231154492

Strydom, P. (2012) Modernity and cosmopolitanism: From a critical social theory perspective. In G. Delanty (ed), *Routledge Handbook of Cosmopolitanism Studies*. London: Routledge, pp 25–37.

Van Dooren, T. (2016) The unwelcome crows: Hospitality in the Anthropocene. *Angelaki*, 21(2), 193–212.

4

Feminist Theories of Justice

Don Mitchell

Introduction

Feminist theories of justice begin, of course, by placing women and gender at the centre of analysis. But they also move well beyond a 'mere' gender-centric view of justice to engage in multidimensional analyses that, on the whole, tend to throw into question many of the basic assumptions of justice theory, whether developed within liberal, libertarian or more radical approaches. This chapter will focus primarily on 'western' feminist theories of justice; further feminist insights are developed in the chapters on postcolonialism (Chapter 6), Indigenous justice (Chapter 7), and environmental justice (Chapter 9) among others.

Key ideas, key theorists

Feminist theories of justice begin – and began, in the work of Olympia de Gouge and Mary Wollstonecraft in the 1790s – with the assertion that women must be understood as individuals with as full and equal status as their male counterparts. This basic assertion was central to the arguments of the women of Seneca Falls and their *Declaration of Sentiments* in 1848, the writings of Harriet Taylor and John Stuart Mill in the 1860s, and the suffrage activists of the 1880s–1920s. The question of women's equal rights reappeared with the revitalisation of the feminist movement in the 1960s and the relationship between rights, justice and the complex politics of identity (race, sexuality, ethnicity, indigeneity, and so on) has been a central focus of feminist justice theorising ever since (Okin, 1989; Williams, 1992).

   Though women must be understood as free and equal individuals, feminists understand individuality to always be conditioned – even formed and made possible – by socially structured, power-infused *relations* with others. A primary relation is that of the family. However, the family cannot be understood as either natural or inevitable and it is not somehow 'beyond justice' (as John Rawls [1971] originally held and some liberals still maintain); nor is it somehow always already *just* (as some tendencies within liberalism and communitarianism suggest). Rather, the family must be understood to be part of the 'basic structure' or society and thus a 'subject of justice' (to use Rawls' terms – terms that, in fact some feminists do not accept [Fraser, 1997]). As Susan Moller Okin (1979; 1989) forcefully argued, the justness of the family must be questioned, and a just family (of whatever configuration) is a precondition for a just society. One reason that some liberals have rejected the family as a subject of justice is that, in their view, family relations are *private* relations, and justice, for them, concerns public relations and institutions. Feminists have countered by throwing into question taken-for-granted assumptions about the relationship between the public and the private, which is also to say the political and the personal. The search for, and analysis of, justice cannot take divisions between public and private for granted: they must be closely examined and their political, social and even geographical preconditions thrown into question.

Neither can 'woman' or 'man' be taken for granted, not only because of the fluidity and constructedness of gender identity but also because it is a question of 'who counts'. At least as far back as Sojourner Truth's famous question, 'Ain't I a woman?', *difference* has been a central factor within women's rights and justice struggles, and *group difference* is particularly important. As Iris Marion Young (1990) argued, group difference is typically accompanied by group-differentiated oppression and domination and any theory of – and struggle for – social justice requires careful attention to these group differences (see Gilmore, 2002). What constitutes any person is determined by dynamics not only of gender, but also race, ethnicity, class, and so forth. In other words, feminist theories of justice require an intersectional analysis (Mohanty, 2003; Crenshaw, 2017).

They also require, as Young (1990, pp 4–5) argued, that concepts of justice cannot 'stand independent of a given social context and yet measure justice', but rather must 'begin from historically specific circumstances, because there is nothing but what is, the given, the situated interest in justice, from which to start' (which is also a point made by Okin [1989]). In this sense, feminist theories of justice tend to be grounded, or materialist, theories. From such a standpoint, which implies 'thinking from women's lives' (Harding, 1988), much feminist analysis argued that a *just distribution* (including within the family) is a necessary, but not even close to sufficient, part of a just society. Rather, as Nancy Fraser (1997; 2008) argues, (economic) *redistributions* need to be considered in relation to (cultural) *recognition* and (political) *representation*. Such recognition demands, in turn and as the Black feminist tradition insists, not only an analysis of how race shapes gender (and vice versa), but also a definite politics of liberation or emancipation (Davis, 1983; Lorde, 1984; James, 1999; Gilmore, 2002). For Joy James (1999), this requires, at minimum, a thorough transcendence of liberal (feminist) traditions and a recovery of the emancipatory potential of radicalism.

Thus, for feminists, a solitary, or even primary, focus on creating – or reforming – the 'basic structure' to create the conditions of possibility for a just distribution (of resources, offices, and so on) is a kind of misdirection. Whatever the subject of justice (and however defined), *injustice* is always domination and oppression, as Young (1990) thoroughly showed. Oppression, Young theorised, has five 'faces' (exploitation, marginalisation, powerlessness, cultural imperialism and violence) and is structural – part and parcel of the structure of society (though it can have interpersonal manifestations). It often arises within contexts in which individuals and groups are behaving morally and ethically in terms of the rules of society and institutions. Justice is therefore not some predefined state, or set of goods, however distributed, but rather arises in efforts to minimise and eliminate domination and oppression, especially, perhaps, when oppression and domination are the result of society otherwise acting lawfully and ethically. In this sense, according to Young (2011), it is profitless to seek to assign guilt for many forms of injustice, and more profitable to find ways to understand *shared responsibility* for producing it and thus shared responsibility for rectifying it.

Gender injustice is a significant aspect of injustice more generally: it is those forms of domination and oppression that are centrally focused on gender, and gender justice is the struggle to eliminate those forms of domination and oppression. Of central importance here, as Raewyn Connell (2011) argues, is understanding how the *patriarchal dividend* – the unearned benefits that accrue to men (or the masculine gender) simply because they are men; that is, the duty regularly required of women by patriarchy – extends well beyond economic and distributional matters.

For contemporary feminist theorists, gender justice and justice more generally are spatially complex, rooted in scales as small as the bedroom and home (or coffee room and office suite) and as large as the globe, and it is the complex interaction of these that must be understood. In these terms, theorising – and struggling for – justice only in relation to the Westphalian nation state is unavailing. Contemporary feminist justice theorising is thus increasingly linked to developed discourses of *global justice*, as in the work of Alison Jaggar (2009), among others. Young's (2011) *social connection model* shows promise for both assessing responsibility for injustice and organising to address it. At the same time, there is a growing recognition of, and efforts to redress, the dominance of feminists from the west/Global North in feminist justice theory and philosophy. Perspectives from the South are necessarily different from, though not necessarily unrelated to perspectives from the North (Connell, 2011). Such work begins from a critical understanding of locally embedded values in relation to histories of colonial domination and oppression (see Chapter 6).

Feminist theories of justice in both the South and the North are, moreover, centrally concerned with *social reproduction*, in a sense reversing the standard Marxist focus on the reproduction of the means of production (and the society this gives rise to) to understand relations of production as seated *within* reproduction. Theorists like Cindi Katz (2001; 2004) and more recently Nancy Fraser (2016) have argued that contemporary capitalist social reproduction is in deep crisis and this crisis reshapes (and inhibits) the conditions of possibilities for justice. Social reproduction is deeply entwined with questions of care, and thus a focus on the *politics of care* is an increasingly prominent and indispensable component of feminist theories and practices of justice, as Virginia Held (2005) insists.

Debates and critiques

There are, of course, significant debates within feminism over how best to conceptualise key aspects of justice, like its substance, the centrality (or not) of distribution, the constitution of procedural justice, and so forth. In broad outline, these debates have unfolded in three steps:

1. The disruptive insertion of gender into justice theorising.
2. The theorisation of gender justice as a central component of the shift from *individuals* to *individuals-in-relation-to-groups* as the subject of justice.
3. The resulting focus on social reproduction, social connection, responsibility and care within the context of the globalisation of justice and the efforts to develop appropriately sophisticated theories of global gender justice.

In turn, these debates have spawned a significant reconsideration of how the main forms of justice – substance, distribution, procedure and retribution – should be understood.

The disruptive insertion of justice

In her indispensable *Women in Western Political Thought*, Susan Moller Okin (1979) shows two things. First, the presumably universal language of 'man' or 'he' is not generic for humanity (and cannot be dismissed as mere dated language). The effect is to exclude women as subjects of politics and philosophy (and thus justice), sometimes actively, as with Rousseau, but more often passively (and thus more consequentially) as with Rawls. Second, and related, when women are recognised as real, individual beings, the questions asked concerning them are different than for men. For men, philosophers ask: 'What are men like?' 'What is man's potential?' But for women, the question is nearly always: 'What is woman *for*?' (Okin, 1979, p 10, emphasis in original). Then, in her later work, Okin (1989) showed a third, exceptionally consequential, thing. When male theorists did start to include women (saying 'men and women' instead of only 'men'), or use more gender-neutral language ('people' instead of 'man'), this tended to be a *false* neutrality that had the effect of further subsuming women into men while appearing to treat each as unique.

The frequently unacknowledged shift from 'what are men like' to 'what are women for' is especially consequential for theories of justice and again for two reasons. First, if (as much liberal and cosmopolitan philosophy hold) a primary basis for a just society is the Kantian imperative that individuals must be treated as ends in themselves and never as means for others' profit, enjoyment or sovereignty, then right at its heart, western philosophy violates one of its most cherished principles. Second, in the western tradition, up to and including the western liberal tradition of justice philosophy (within which Okin places herself), what appears to be about individuals in a polity is really about the *patriarchal family* in society. Women are subordinated – actively – and made to exist insofar as they are *for* their husbands and fathers. The assumption of individuality is always violated (Okin, 1989, p 202). Indeed, Rawls (1971) is inadvertently explicit about this; those in the 'original position' in his theory are 'heads of families', not 'individuals' as such (a position he did not revise until quite late in his life).

Okin (1989) developed this analysis with devastating effect on much justice theorising, especially within the liberal, libertarian and communitarian traditions. Though there are exceptions among male philosophers, most prominently J.S. Mill, who argued in *The Subjection of Women* (1869) that '[j]ust treatment no less than liberty is [to be] regarded as essential for the happiness of women themselves and as a necessary condition for the advancement of humanity' (Okin, 1989, p 214), western philosophy assumes that the family is 'beyond justice', or always-already just. Mill differed, arguing that families were frequently unjust and that unjust families were 'a school of despotism'; but he also never really questioned the division of labour within the family (Okin, 1989, pp 20–21). Yet, Okin (1989, p 4) argued, a just society would only be possible if it was rooted in just families (however configured), families – or households – in which a just division of labour, rather than an exploitative one, obtained: 'Until there is justice within the family, women will not be able to gain equality in politics, at work, or in any other sphere.'

From individuals to individuals-in-relation-to-groups

Taking gender seriously, Okin made clear, requires a thorough reconceptualisation of both the subject and object of justice and therefore a transformation of the *most basic* structures of society, like the family. Okin is careful in her analysis not to assume or over-valorise the heterosexual, nuclear family, nor does she assume gender as given, but even so, her work sits reasonably comfortably within White, western, liberal thought. By the time she published *Justice, Gender, and the Family* (1989), however, the assumption of monadic individuality that undergirds her work was under multi-flanked attack.

Black feminist theories and theorising from the 'third world' together contributed to a questioning of feminism's understanding of the liberal (gendered) subject. Radical, emancipatory work by scholars and activists like Angela Davis (1983) and Joy James (1998; 1999), the influential arguments related to identity – and social struggle – by the Combahee River Collective (Taylor, 2017) including Audre Lorde (1984; 2017) and bell hooks (1984; 2000), the work of borderlands scholars like Gloria Anzaldúa (1987), and the reconceptualisation of feminism as a decolonising project by Chandra Mohanty (2003), among others, all contributed to 'dethroning' the liberal, White (middle-class) woman as the subject of feminism and the development of a much more complex politics of identity – as formed through interlocking systems of oppression – in the last quarter of the 20th century. Eventually codified by Kimberlé Crenshaw (2017) as 'intersectionality', this reorientation of feminist theory is profoundly important for theorising justice. Remarkably, however, there has been surprisingly little direct engagement by feminists of colour with theories and philosophies of justice. While the struggle for social justice is frequently invoked (and, importantly, such justice struggles have been the focus of significant historical and sociological work), justice itself is rarely theorised. It is typically assumed to be the opposite of oppression.

Even so, the importance of intersectionality and group difference (and the relations of power that structure these) has been directly incorporated into some feminist justice theorising, even as aspects of it have been contested. Fraser (1997, p 1), for example, worried that concern with group identity was coming to supplant 'class interest as the chief medium of political mobilization' and particularly the development of a situation – in the world as well as in theory – where 'cultural domination supplants exploitation as the fundamental injustice' and she thus argued for a theory of justice that understood recognition in relation to redistribution. Jaggar (2009, pp 5–6) saw matters differently. For her, critical race theory (among others) helped expand the *domain* of justice (not just supplant one domain with another) to incorporate institutions as well as individuals, the *subjects* of justice to include groups as well as individuals, and the *objects* of justice to include recognition as well as redistribution. None of this is that far from Fraser's formulations (especially Fraser, 2008), but the weight of emphasis is different.

Social reproduction, social connection, responsibility and care: global justice

As a consequence of these debates, feminism demands that the domain of justice includes the personal (or domestic, in Okin's terms) and that the object of justice includes responsibility for unpaid care-work, and thus that matters of *social reproduction* take centre stage (Katz, 2001; 2004; Fraser, 2016). As Katz in particular showed, gendered structures of social reproduction are under severe strain in the post-Fordist, neoliberal, globalising era (which is also the postcolonial era). Under such pressures, feminists have sought to construct a more global feminism, a 'feminism without borders' in Mohanty's (2003) phrase, that has had profound effects on feminist justice theorising. Young's (2011) *social connection model* is designed to understand responsibility in a globalised world in which, for example, exploitative supply chains are global in scope (as will be discussed in the section 'Procedure and distribution', later).

For her part, Jaggar (2009) argues that globalisation and its colonial/ postcolonial legacies, together with the increasing importance of international law, has shifted nation states from being the *domain* of justice to being *subjects* of justice, at least in part. For this reason, she argues, there is a need to further develop notions of – and possibilities for – a 'global basic structure'. At the same time, the *objects* of justice are now dispersed or distributed as much as, or more than, they are consolidated (for example in the nation state, city or home). Analyses of gendered justice in the contemporary conjuncture thus cannot afford to be particularistic, but must focus on the extended and distributed networks of power and processes that shape the contemporary world. As regards *principles* – how justice should be enacted – this also requires considerable reconsideration of the structures of power that guide political intervention and new modes of solidarity (a key concern of Young at the end of her life, and a central focus on the work of Ann Ferguson [2009]).

Substance and distribution

As already indicated, the earliest feminist theorising (as well as organising and struggle) in relation to justice concerned women's inclusion in the polity, their own standing as full human beings and thus subjects of justice in their own right, and not either subsumed by, or understood to be appendages of, men. This is, obviously, the most fundamental *substantive* justice question: Who has the right to have rights (as Hannah Arendt [1951] influentially framed the matter)? Who counts? Who has a recognised voice? These questions remain vital today. They have been expanded in one direction to question the necessity of *humanity* (can other species, or whole ecosystems, count?), in a second to question temporality (do past and future generations count?), and in a third to question scale – or the 'where' of justice (is the family or household a subject of justice? Is the nation state the appropriate 'container' for justice claims, especially given the facts of globalisation?).

Such questions lead to a central debate within feminism: to what extent can we theorise a *universality* of justice? Martha Nussbaum (1997) defends liberalism and its universalising theories of justice from critics like Alison Jaggar (1983, pp 47–48), who argue that 'the liberal conception of human nature and political philosophy cannot constitute a philosophical foundation for an adequate theory of women's liberation' because of its stark individualism and essentialising tendencies. For Nussbaum, liberalism's core assumption of individualism is precisely its strength:

> Liberalism does think that the core of rational and moral personhood is something that all human beings share, shaped though it may be in different ways by their differing social circumstances. And it does give this core a special salience in political thought, defining the public realm in terms of it, purposely refusing the same salience in the public political conception to differences of gender and rank and class and religion. (Nussbaum, 1997, p 23)

If there is to be a feminist critique of liberal universalism and individualism, according to Nussbaum (1997, p 13), then it is simply that liberalism has not been individualist enough.

Susan Moller Okin (1989) made something of a similar argument in her examinations of how women's individuality – and humanity – has been subsumed into the family and subordinated to male authority, both in liberal philosophy and in historical practice. When, in this view, the universal fact of women's selfhood is truly taken seriously, as Nussbaum (1997, p 2) argued was beginning to be the case in international and human rights law in the 1990s, then and only then would liberalism's 'radical feminist potential ... [begin] to be realized'. On this account, substantive gender justice will be accomplished when and to the degree women's personhood comes to be accepted, protected and legally defined. Substantive justice, to put it somewhat oversimply, inheres in the degree to which women's personhood is not violated, socially, politically, in law, through violence, or otherwise.

It also inheres in another, empirical, fact. When, as Okin (1989) argues, standard theories of justice assume a false gender-blindness, feigning a world in which gender equality already exists, they inevitably imply the *actual* oppression of women. To the degree, for example, that there remain unequal and unjust divisions of labour in the home, granting women equal status by theoretical fiat, without also attending to these domestic power relations, means that women have to be free and equal citizens *and* tend the home and family while men only need to be free and equal citizens. Assuming a false equality in these cases leads to theories, and thus likely to policies, that further oppress women. Substantive justice, taking the form of full equality and sovereign individuality, leads to substantive injustice in the form of this double burden. And it is for exactly this reason that Okin insists that the family is and must be a 'subject of justice': if the family is set off-limits as part of the private sphere (as in Rawls, 1971), there can be no chance for justice.

Feminist critics of liberal feminism take this argument a step further. They do not deny the centrality of women's personhood or the marginalisation and violence that accompanies women's subsumption into men and their interests (that is, the fundamental injustice of defining women *for* men or *for* the family, as Okin [1979] describes). But (as noted) they often work from a rather different theory of personhood, and this has important implications for understanding the substance of justice. Rather than a monadic ontology of personhood, more radical feminist philosophers understand that the *individual* is *indivisible* (the worlds are closely related [Williams, 1983, pp 161–165]) from the groups of which she is a part: individuality is *relationally* determined. This is to say that there are no individuals without groups to give them form and, indeed, individuality. Understood in this way, close attention to group differentiation in theories of justice is as vital as close attention to individuals as subjects of justice. As Young (1990, p 43) puts it, '[s]ocial groups … are not simply collections of people, for they are more fundamentally intertwined with the identities of the people described as belonging to them. They are a specific kind of collectivity, with specific consequences for how people understand one another and themselves'. The individual does not pre-exist the group. In Young's (1990, p 43) words again, '[p]olitical philosophy typically has no place for a specific concept of the social group. When philosophers and political theorists discuss groups, they tend to conceive of them on the model of aggregates or the model of associations, both of which are methodologically individualist'. In this view, it is insufficient to focus theorising on individuals *qua* individuals (as liberalism does). Instead, a more multidimensional analysis is required that understands individuals in relation to groups and each other.

Socialist-feminist theorising, such as that associated with Fraser, Jaggar and Young, starts from just this ontological assumption, and the implications for how such feminist approaches understand the substance of justice, and how it differs from liberalism, are clear. First, *justice substantively concerns the just treatment of individuals, but only insofar as that treatment is just for the group.* A just distribution of goods, offices and opportunities (the traditional focus of liberal justice theorising) remains a vital focus in radical feminist philosophies of justice, but such redistribution must be analysed in relation to questions of – and claims for – *recognition*, often precisely of ontologically essential group differentiations that have been and are marginalised and silenced. In this sense, redistribution and recognition cannot be divorced; each must be predicated on the other.

Second, as Okin (1989) and nearly all other feminist philosophers of justice argue, since theories of justice must not be distracted by abstractions and ideals, but must concern themselves with what actually exists, *the substance of justice inheres not in some ideal, but arises from within, and is defined by struggles against actually-existing injustice.* Young, for instance, argues that injustice is domination and oppression (maldistribution is a function of these, more than vice versa) and something is substantively just when it undercuts, ameliorates or eliminates group-enabled and group-defined domination and oppression (and thus their particular effects on group-defined individuals).

Third, therefore, *the substance of gender justice is that which counteracts gender-based domination and oppression while not enhancing* (indeed while seeking to counteract) *domination and oppression operating through other group differences.*

Finally, then, this sort of radical feminist philosophy is also radically anti-essentialist in that it understands all factors of group differentiation (gender, class, sexuality, race, and so on) as historically and socially produced, no more pre-given than human individuality. In this sense, *the substance of justice also therefore inheres in power – in this case the 'power to define'* (cf Western, 1981). A just society is one in which all members, within and because of their group differentiations, have access to the power to define – in collaboration, in struggle and in full relation to the groups of which they are a part – the conditions of their being, not as monadic individuals, but as fully social beings.

Procedure and distribution

Much of Okin's work was devoted to the critiquing and repairing of the standard mid-20th-century canon – the world of Rawls, Sandel, Nozick, Dworkin, Walzer – by showing what happens when women are not ignored, not subsumed into men, or not falsely included through bogus gender-neutral language, which enabled her to develop some important tenets for understanding what is and is not procedurally just. At the most basic level, a practice cannot be procedurally just if it subsumes the interests of one individual into the interests of another. To speak of 'heads of families' (as Rawls [1971] did in his most influential work) already indicates that a theory of justice will lead to procedurally unjust outcomes, whatever its other virtues (Okin, 1989).

This is, of course, a question of *recognition*, or as legal scholars put it, a question of who has 'standing': who has the right to participate of their own accord in some process, practice or institution. As a whole, feminist theories of justice are centrally concerned with this question of recognition (Fraser and Honneth, 2003; Fraser, 2008). But equally important, in relation to procedural justice, is who has 'voice' – who has the ability to be heard and have their concerns addressed (Fricker, 2007). This matter of *inclusion* (Young, 2000) is of vital importance given the shifting scales at which (in)justice operates. The nation state can no longer in any simple sense be understood as the natural container for justice. A vital question from Fraser thus becomes the procedures by which representation becomes possible and a reality. 'Rethinking the public sphere' (Fraser, 1991) requires rethinking the *scale* of the public sphere and the institutional structures that can produce new scales of representation.

For Young (2000; 2011), these questions led in a somewhat different direction and addressing them entailed a significant shift in theorising people's relation to processes that produce injustice. Much justice theory, and most law, seeks to attribute guilt and culpability for the creation of some wrong. Young calls this a 'liability model'. It is a model that asks who is liable for the creation of some condition and then seeks redress from them. Young did not deny the importance of assigning liability, but recognised its limitations. In particular, she argued that many processes, relations and practices may be perfectly just – moral, ethical and following the rule of law – and still lead to unjust outcomes. Under these circumstances, 'no one' is to blame. Yet all who are implicated in the processes nonetheless bear some *responsibility* for the outcomes. Guilt, according to Young, is backward-looking and therefore not necessarily oriented towards more just futures; the assignment of guilt might do nothing to transform putatively just institutions, systems, and so forth that produce injustice. Responsibility is forward-looking. When we take responsibility for the production of injustice, we seek to transform the conditions that produce injustice so they stop doing so.

Incomplete at the time of her death, Young's arguments concerning how responsibility can be discharged in a solidaristic manner (her 'social connection model') are compelling, perhaps most importantly for the clarity with which they show the inadequacy of theories – like much of the Rawlsian tradition – aiming simply to get procedures right. Though compelling, her arguments have not been convincing to everyone. In her introduction to Young's (2011) posthumous *Responsibility for Justice*, for example, Martha Nussbaum presents an equally compelling critique of Young's divisions between backward- and forward-looking approaches, noting that the refusal to look back means that we always start from an imminent present. If we delay acting today, we are absolved from guilt as long as we act tomorrow. But tomorrow, we are once again absolved as long as we act the next day. Young's theory suffers, perhaps, from infinite regress. But it does not have to, at least not if a fuller theory of responsibility is developed than Young was able to achieve in her lifetime.

Such a fuller theory might begin, following Fraser (1997), by assessing any interventions into process – today *and* tomorrow – in relation to their *affirmative characteristics* and *transformative potential*. Affirmative interventions tend to ameliorate a wound but, at minimum, leave the injuring processes untouched and more likely prop them up and reinforce the status quo.

Affirmative interventions are exactly the 'charity' Mary Wollstonecraft (1792 [1988]) railed against 230 years ago when she declared: 'It is justice, not charity, that is wanting in the world'! By contrast, potentially transformative interventions serve to transform basic conditions, the 'basic structure'. Taking responsibility, as Young wanted us to do, requires identification of potentially transformative interventions and working towards solidaristically implementing them.

Conclusion

Taken together, feminist theories offer a set of key propositions that are indispensable for social scientists seeking to understand the constitution of, and the struggle for, justice. They not only force serious consideration of the who, what, where, when and how of justice (Jaggar, 2009), but in doing so they require a reconsideration of the individualism that undergirds much liberal – and common sense – thinking about justice. They contest methodological, but also ontological, individualism in social science research in general and research on social justice in particular. They require that any focus on (re)distribution as a core of justice must be understood in relation to recognition and representation. Research on redistributive practices and policies that fails to consider effects of and on recognition and representation is simply inadequate. They require, therefore, taking intersectionality seriously. Intersectionality is ontological, since individuals exist only insofar as they form dialectically with and as part of groups, and therefore not merely epistemological or methodological. Finally, feminist theories turn social science work in the direction of understanding the complex, distributed nature of responsibility. To put this schematically, feminist theories of justice turn attention towards:

- gender, but only in relation to
- other factors of identity with which gender is enmeshed, which always implicates
- geographical scale, or a reassessment of the *where* of justice, which requires a particular focus on
- the family/household, as well as
- the global, in all its unevenness, and
- the scales in between.

More specifically feminist justice theories ask social scientists to consider:

- How justice only appears in the struggle to address questions of domination and oppression, that is, *injustice*.
- How domination and oppression operate through complex group differentiation and thus, since individuals are defined through their

membership in and indivisibility from groups, any intervention will have uneven effects depending on 'position'.

- How injustice is structural and not (or not only) interpersonal or epiphenomenal.
- How policies, practices and interventions will always potentially either entrench or ameliorate injustice *within* families and households – the private sphere cannot be ignored and neither can the structures and practices of social reproduction.
- How policies, practices and interventions will always have effects that extend across nation-state (and other governance) lines, and may have grossly uneven, gendered effects in different, seemingly disconnected locales, given the grossly uneven development of geographical space.
- And finally, therefore, how or whether policies, practices and interventions are likely to be *affirmative* of the status quo or *transformative* of it, and if the latter, how it is the direction of change (towards or away from justice, towards or away from enhancing the 'patriarchal dividend') that matters.

Of central importance for social science research, then, is that feminism does not define what is *just* a priori, but understands justice as a (potentially transformative) move away from domination and oppression. The content of justice arises in, and is internally related to, this move. The content of justice is a function of responsibility, care and social connection.

References

Anzaldúa, G. (1987) *Borderlands/La Frontera: The New Mestiza*. San Francisco, CA: Aunt Lute Books.

Arendt, H. (1951) *The Origins of Totalitarianism*. New York: Harcourt, Brace and Co.

Connell, R. (2011) Gender and social justice: Southern perspectives. *South African Review of Sociology*, 42(3), 103–115.

Crenshaw, K. (2017) *On Intersectionality: Essential Writings of Kimberlé Crenshaw*. New York: New Press.

Davis, A.Y. (1983) *Women, Race, and Class*. New York: Vintage.

Ferguson, A. (2009) Iris Young, global responsibility, and solidarity. In A. Ferguson and M. Nagel (eds), *Dancing with Iris: The Philosophy of Iris Marion Young*. Oxford, UK: Oxford University Press, pp 185–197.

Fraser, N. (1991) Rethinking the public sphere: A contribution to the critique of actually existing democracy. In C. Calhoun (ed), *Habermas and the Public Sphere*. Cambridge, MA: MIT Press, pp 109–142.

Fraser, N. (1997) *Justice Interruptus: Reflections on the 'Postsocialist' Condition*. New York: Routledge.

Fraser, N. (2008) *Scales of Justice: Reimagining Political Space in a Globalizing World*. Cambridge, UK: Polity.

Fraser, N. (2016) Contradictions of capitalism and care. *New Left Review*, 100, 99–117.

Fraser, N. and Honneth, A. (2003) *Redistribution or Recognition: A Political-Philosophical Exchange*. London: Verso.

Fricker, M. (2007) *Epistemic Injustice: Power and the Ethics of Knowing*. New York: Oxford University Press.

Gilmore, R.W. (2002) Fatal couplings of power and difference: Notes on racism and geography. *Professional Geographer*, 54(1), 15–24.

Harding, S. (1988) *Whose Science? Whose Knowledge? Thinking from Women's Lives*. Ithaca, NY: Cornell University Press.

Held, V. (2005) *The Ethics of Care: Personal, Political, and Global*. Oxford, UK: Oxford University Press.

hooks, b. (1984) *Feminist Theory: From Margin to Center*. London: Pluto.

hooks, b. (2000) *Where We Stand: Class Matters*. New York: Routledge.

Jaggar, A. (1983) *Feminist Politics and Human Nature*. Totowa, NJ: Rowman and Allanheld.

Jaggar, A. (2009) The philosophical challenges of global gender justice. *Philosophical Topics*, 37(2), 1–15.

James, J. (ed) (1998) *The Angela Y. Davis Reader*. Oxford, UK: Blackwell.

James, J. (1999) *Shadow Boxing: Representation of Black Feminist Politics*. New York: St. Martin's Press.

Katz, C. (2001) Vagabond capitalism and the necessity of social reproduction. *Antipode*, 33(4), 709–728.

Katz, C. (2004) *Growing Up Global*. Minneapolis, IL: University of Minnesota Press.

Lorde, A. (1984) *Sister Outsider: Essays and Speeches*. New York: The Crossing Press.

Lorde, A. (2017) *Your Silence Will Not Protect You: Essays and Poems*. London: Silver Press.

Mill, J.S. (1869) *The Subjection of Women*. London: Longmans, Green, Reader and Dyer.

Mohanty, C.T. (2003) *Feminism without Borders: Decolonizing Theory, Practicing Solidarity*. Durham, NC: Duke University Press.

Nussbaum, M. (1997) The feminist critique of liberalism. The Lindsay Lecture, University of Kansas, 4 March. https://kuscholarworks.ku.edu/bitstream/handle/1808/12410/The%20Feminist%20Critique%20of%20Liberalism-1997.pdf?sequence=1.

Okin, S.M. (1979) *Women in Western Political Thought*. London: Virago.

Okin, S.M. (1989) *Justice, Gender, and the Family*. New York: Basic Books.

Rawls, J. (1971) *A Theory of Justice*. Cambridge, MA: Harvard University Press.

Taylor, K.-Y. (ed) (2017) *How We Get Free: Black Feminism and the Combahee River Collective*. New York: Haymarket.

Western, J. (1981) *Outcast Cape Town*. Berkeley: University of California Press.

Williams, P. (1992) *The Alchemy of Race and Rights*. Cambridge, MA: Harvard University Press.

Williams, R. (1983) *Keywords*. London: Fontana.

Wollstonecraft, M. (1988 [1792]) *A Vindication of the Rights of Woman*, edited by C.H. Poston. New York: W.W. Norton.

Young, I.M. (1990) *Justice and the Politics of Difference*. Princeton, NJ: Princeton University Press.

Young, I.M. (2000) *Inclusion and Democracy*. New York: Oxford University Press.

Young, I.M. (2011) *Responsibility for Justice*. New York: Oxford University Press.

5

Radical Justice: Anarchism, Utopian Socialism, Marxism and Critical Theory

Don Mitchell and Johanna Ohlsson

Introduction

Radical accounts of justice are less a coherent normative tradition or school of philosophy than a set of positions on or orientations towards justice, typically grounded in a critique of the *in*justice of contemporary social arrangements and their need for transformation. To be 'radical' is to seek to affect 'the fundamental nature of something', to be 'innovative or progressive', to offer diagnoses and interventions 'intended to be completely curative', and to advocate 'thorough political or social reform' often through measures thought to be 'politically extreme' at least in relation to mainstream politics, to adopt some of the definitions offered by the *Shorter Oxford English Dictionary*. In this sense, radical approaches to justice often set themselves in opposition to liberal theories, which, from a radical standpoint, are understood to excuse, even support, an unjust status quo. Such a critique of liberalism (which may nonetheless adopt some of its normative tenets) provides a common denominator for radical theories of justice that are otherwise quite diverse in scope. This chapter surveys anarchist, utopian socialist and Marxist approaches to social justice from their foundation in the 19th century to their elaboration within two influential centres of western Marxist thinking that have proved to be especially influential in the social sciences: the Frankfurt School of critical social theory and the spatialisation of Marxist thought in the work of David Harvey (among others). We conclude by briefly noting the implications of Marxian/socialist thinking about justice for the 'aspects of justice' identified in the introduction to this book: the who, what, where, when and why of justice.

Key ideas: debates and critiques

Most modern radical theories of justice have their origins in the Enlightenment and especially in the reaction to the expansive, radically transformative evolution and effects of the spread of capitalism from its early modern mercantile colonial form to its later modern industrial imperialist form. Such historical development was radically uneven and contradictory and often retained strong elements of its feudal precursor, deepening rather than ameliorating inequalities in modernising societies. In this context, the demand for liberty, equality and solidarity – now taken to be central tenets of a just, liberal society – were, of course, extremely radical in the context of the French Revolution, as was the contemporaneous demand by Olympia de Gouge and Mary Wollstonecraft that women be afforded full and equal standing in society (see Chapter 4). Together with the Haitian and American Revolutions, the French Revolution helped ensure that the end of the 18th and first half of the 19th centuries would be an 'age of revolution', especially in Europe (Hobsbawm, 1962).

Radical justice in utopian socialism and anarchism

Utopian socialists like Charles Fourier (1772–1837) and to a lesser extent Robert Owen (1771–1858) sought to imagine, and create, the social and spatial preconditions for a society of radical egalitarianism, which for them was the primary precondition for a just society. For Fourier, the emancipation of women was a basic measure of the progress of society towards social justness. It also required sexual liberation (including openness to sexual diversity) and the reassociation of work with libidinal pleasure (see, for instance, Marcuse, 1955). Work was to be organised cooperatively with less desirable occupations more highly compensated and the resulting social product distributed according to need. A just society was a pleasurable society and one that allowed people, through their cooperative and libidinal endeavours, to liberate their human passions. Fourier's ideas were foundational for the efflorescence of Utopian communes and intentional communities that spread across North America and Europe in the 19th century and have continued to be inspirational to socialist and utopian thinkers (Hayden, 1976; Harvey, 2000).

Pierre-Joseph Proudhon (1809–1865) is reputed to be the first to use the term 'anarchist'. Like Fourier, he was possessed of an antisemitic temperament, but unlike Fourier, he was also deeply anti-feminist, defending patriarchy to the hilt, a position quite in conflict with many of his anarchist ideas, especially his central argument that justice had to be founded on equity. His brand of anarchism was based on mutualism, which was likewise central to the anarchism of Élisée Reclus and Pyotr Kropotkin (2006 [1892]; 1902). In particular, Proudhon advocated what he called 'industrial democracy' where workers, organised in labour associations, freely cooperated and exchanged their products on the market with other cooperative labour associations. By contrast, Reclus and Kropotkin's version of 'mutual aid' was based more on the 'needs principle' ('from each according to their abilities to each according to their needs') than market principles. In Proudhon's world, wages – and thus labour power as a commodity – would be abolished, as would be the state, which would be replaced by federations of free communes (municipalities). Famously associated with the phrase 'property is theft', Proudhon's views on property were complex (and evolved over his lifetime), but he essentially held that 'property' in personal goods was acceptable, but monopoly ownership of land and the means of production, especially when used as a means of labour exploitation, was not.

Proudhon's view of human emancipation was primarily confined to emancipation from being governed, arguing that the 'justice' and 'morality' of being governed consisted only in being condemned, judged, ridiculed, spied on, exploited and oppressed (Proudhon, 1923 [1851]). Anarchist justice thus consisted in the opposite of these, and in liberation from them through mutualism. Such ideas – implicating just systems of social production through cooperation, just distributions of social products through social ownership, and just governance through liberation from state tyranny – have remained cornerstones of anarchist thought ever since, as has been well summarised by David Wieck:

> A society will be just, then, insofar as it is free … of 'enslaving' social or political institutions (military, familial, governmental, educational, sexual, ethnic hierarchical, ecclesiastical, etc.); but it will not be a society at all unless patterns of cooperation capable of sustaining human communities and vital personal existence are achieved. (To be anarchist and just, a society need not be perfectly, or even approximately egalitarian in an economic sense, unless such a principle arises from mutual agreement; unjust would be such systematic discrepancies of wealth as would constitute de facto economic classes, where the inferior class or classes would be chronically blocked off from full participation in the life of the society.) (Wieck, 1978, p 231)

Marx and Engels

In his early works and reflecting the influence of Hegel (and to some degree Rousseau) on his thinking, Karl Marx was essentially concerned with the question of human flourishing, seeking to understand what would allow and what would thwart humans from achieving their 'species being' – their potential as humans. In 'On the Jewish question', Marx (1844) made a distinction between 'political emancipation' (that is the granting of full citizenship to Jews) and 'human emancipation' (in this case emancipation from the mystifications of religion). In the *Economic and Philosophical Manuscripts* (1932 [1844]), he sought to account for the forces producing humans' alienation from their own nature, focusing on the inhibiting, alienating and exploiting effects of the capitalist mode of production and arguing that under capitalism, humans are alienated (estranged) from:

- what they produce;
- their own labour;
- other human beings;
- their own selves; their own natures.

Finally, in *The German Ideology*, Marx and Engels (1932 [1846]) show that what differentiated humans from other animals, and thus what is central to their nature, their species being, is that they can and must produce their own means of subsistence. What a human *is* – and *can be* – is determined by the material circumstances of such production. Significantly, production is always and must be social in character. It is through collaborative acts of production that humans produce themselves and their actually-existing nature, however alienated, and however far removed from their species being.

Though Marx rarely invoked *justice* in his writing, and though he was sceptical of rights in the abstract (though less so as actual social practices), his theory of alienation, especially when combined with his theory of exploitation, has profound implications for theories of justice. Central to Marx's arguments – especially in later writings like the *Grundrisse* (1976 [1856]) and *Capital* (1987 [1867]) – was that under capitalism, there is nothing unjust about exploitation. Exploitation (which is necessary to the production of surplus value and thus the accumulation of capital – or 'economic growth' in mainstream parlance), is the result of formally *just* market interactions whereby a worker sells her/his labour power at its value and for a mutually agreed upon length of time. The purchaser of that labour power (the capitalist) is thus free to deploy that labour as seen fit (within the laws and mores of the land). That the workers can produce sufficient commodities to repay her/his value (as represented in the wage) in less than the agreed time, and thus continue to work for the capitalist 'for free', is exactly what makes the system go. That the product of all that expended labour power now belongs solely to the capitalist is juridically as it should be. For Marx, 'exploitation' was a measurement of the relationship between a worker's 'necessary' labour time (the time required to replace her/his own wages) and his/her 'surplus' labour time (the time spent working 'for free'). Exploitation was a technical term. But exploitation in practice was exactly the site of human alienation and the thwarting of our species being: by selling our labour power ('justly') we alienate ourselves from what we produce (this is determined by the capitalist), from how we produce it (ditto), from each other (all our relationships are mediated through and take the form of commodities), and from ourselves (we have little or no control over our own self-development) (Geras, 1985).

If what is *just* within capitalism is deemed more broadly as *unjust* (because it thwarts human species being, or more narrowly because despite its juridical legitimacy it creates gaping inequalities in income, life chances, self-development, and so forth), then, as with anarchism and the ideas of the utopian socialists, the only way to create a more just society is to radically transform the current one, especially, for Marx, in how it organises the social relations of production. More analytically, the full corpus of Marx's work suggests a dual nature to justice:

1. The substance and processes of justice are historically determined and geographically situated: in one era and society, slavery, for example, might be perfectly just (as in Greek and Roman worlds), bonded labour might be just (as in feudalism), or child labour might be just (as in much of the world, including the industrial capitalist world, into the 20th century). As an important corollary, standards of justice change only as the result of concerted, and often long-term, social struggle.
2. Nonetheless, according to Engels, justice is an ideal to be strived towards as well as a 'stick' against which to measure current society – 'the final arbiter to be appealed to in all conflicts' (quoted in Merrifield and Swyngedouw, 1995, p 1). This 'stick' is human emancipation, the degree to which humans achieve their 'species being', which is to say, humanity's full potential as a species.

The complex dialectic between the material (justice is historically and geographically determined) and the ideal (justice is the stick against which to measure current society) suffuses all of Marx's writing, but given his foundational historical materialist argument that it is social life that determines consciousness, not consciousness that determines social life, his analytical work was more concerned with understanding the logic of capitalism (as the dominant force shaping humans' *actual* human nature) than with theorising justice as such (see Forst, 2017, p 113; Wolff, 2017). For this reason, perhaps, it is easier to spot critiques and theories of *injustice* in his work than it is to find sustained discussions of justice, which is why the preceding focus on the overall thrust of his work has been important.

The Frankfurt School and Critical Theory

By contrast, justice is more explicitly theorised in the work of the Marxian Frankfurt School, especially after its post–Second World War revival. The first (pre-war) generation of Frankfurt theorists returned Marxism to its Hegelian roots (while juicing it up with a good dose of Freud) and thus placed an emphasis on *critique* and *theory*. Max Horkheimer (1972, p 246) conceived of Critical Theory as a theory that contributed to human 'emancipation from slavery' and helped 'to create a world which satisfies the needs and powers' of human beings. As with Marx (and the anarchists), creation of a just world requires a thorough remaking of society. This in turn requires the critical examination of the forces, including the ideological forces, that can lead institutions to create the conditions of possibility for freedom and justice (see Held, 1980). Theodor Adorno and Horkheimer together launched a withering critique of the failures of the Enlightenment to live up to its own ideals, with Adorno arguing (in Schick's [2009, p 147] words) that 'Enlightenment notions of justice and injustice fail to live up to their goal of improving well-being'. Furthermore, the appearance of progress towards Enlightenment ideals hinders critique: 'the semblance of freedom makes reflection upon one's own unfreedom incomparably more difficult than formerly' (Adorno, 1981, p 21).

For this reason *justice* is closely linked to *ideology*. For early Frankfurt School theorists, ideology:

> is *justification*. It presupposes the experience of a societal condition which has already become problematic and therefore requires a defense just as much as does the idea of justice itself, which would not exist without such necessity for apologetics and which has as its model the exchange of things which are comparable. (Adorno, 1972 [1954], pp 189–190, emphasis in original)

For the first generation of Frankfurt theorists, to the degree that Critical Theory is ideology critique, then Critical Theory's contribution to justice theory is its critique of justice *as* ideology, as 'apologetics' for the status quo. Beyond an orientation towards human emancipation and a methodology of immanent critique aimed at exposing the dialectical underside of modernity and Enlightenment, however, this generation of Frankfurt theorists made little contribution to a *positive* theory of justice, that is, a theory that moves beyond critique to develop a rational basis for what is just.

The same cannot be said for later (post-war) generations. There is no satisfactory way to summarise the range and depth of Jürgen Habermas' development of Critical Theory over his long career, nor even the range of its implications for justice theory. But at the risk of oversimplification, it can plausibly be argued that Habermas has been consistent in attempting, from *The Structural Transformation of the Public Sphere* (1962), through his major work on communicative rationality (1981), to his more recent works on cosmopolitanism and democracy (2001; 2012), to develop what Pettit

(1982, p 228) has identified as a 'consensus theory of justice'. Such a theory, clearly echoing Horkheimer but with less negative connotations, is rooted in a theory of rational *justification*. For Habermas, justice, like truth, is a 'discursively resoluble validity claim' (quoted in Pettit, 1982, p 228), which is to say it can more fully be empirically investigated than normatively defined. And yet, into the 1980s, according to Pettit (1982, p 228), there *was* a normative core to Habermas' sense of justice: a 'just system is that which impartially and maximally satisfies people's real needs' – which is to say an essentially distributive form of justice arrived at through maximally rational procedures of discursive engagement. In later work, particularly *Between Facts and Norms* (1992), and in a series of debates with John Rawls, Habermas developed a 'critical theory of justice' that was discursively grounded (as it also is with Rawls) on the argument that 'justice itself has no authority other than that which it "earns" in a justified way; public justification remains the "touchstone" of normativity' (Forst, 2014, p 156, see also Forst 2010).

Like Rawls, Habermas is centrally concerned with the 'basic structure' – the institutional arrangements within which we all must live – as the primary object of justice, but 'presupposes a model in which the citizens accept the conception of justice based on publicly sharable reasons, such that an actual moral consensus independent of comprehensive doctrines exists' (Forst, 2014, p 163). Yet Habermas does not deny that there is a foundational content to justice. As he says in perhaps his most straightforward statement: 'Justice concerns the equal freedoms of unique and self-determining individuals' (Habermas, 1990, p 244). Such content must be understood in a particular way: as the 'reverse side', that is, indissoluble from, *solidarity*. As a fuller rendering of Habermas' statement puts it:

> Every autonomous morality has to serve two purposes at once: it brings to bear the inviolability of socialized individuals requiring equal treatment and thereby equal respect for the dignity of each one; and it protects intersubjective relationships of mutual recognition requiring solidarity of individual members of a community, in which they have been socialized. Justice concerns the equal freedoms of unique and self-determining individuals, while solidarity concerns the welfare of consociates who are intimately linked in an instersubjectively shared form of life – and thus also to the maintenance and integrity of this form of life itself. (Habermas, 1990, p 244)

Within this context of solidarity, *justice* must be publicly defended, justified, and it is only in its justification that justice takes on real, practical meaning. And yet, recently, Habermas (2014; see Peirce, 2017) has reversed himself on this position and now argues that solidarity is *not* an essential aspect – the reverse side – of justice, moving closer to Rawlsian liberalism, and thereby, perhaps, diminishing the radical core of his theory of justice.

Habermas' Frankfurt colleague, Rainer Forst (2012), has sought to retain that radical core. As with Habermas, justification is the touchstone. For Forst (2014; 2017) justice is *non-domination* and the *right to justification*, which is operationalised through *reciprocity* and *generality*, respectively. Reciprocity means that 'one does not make any claims to certain rights or resources that one denies to others, and that ones does not project one's own reasons (values, interests, needs) onto others in arguing for one's claims' (Forst, 2004, p 317). Generality means that all affected persons must be able to access and accept the reasons (for a claim of justice) in relation to universal and fundamental norms (Forst, 2012, p 6).

These base arguments are linked to a reorientation of justice theory from 'recipient-oriented views' to '*production* and its just organization' (Forst, 2017, p 122, emphasis in original), which is the radical kernel of Forst's theory. He argues that most mainstream theories of justice are distributive and 'understand "distributive justice" exclusively as a matter of allocating goods' (Forst, 2017, p 122). Such theories:

- 'obscure essential aspects of justice – in the first place how the goods to be distributed come into the world';
- 'neglect the *political* question of who determines the structures of production and distribution and in what ways – hence the question of power – as if there could be a giant distribution machine that only needs to be program correctly' (cf Young, 1990);
- 'disregard … [the fact] that *justified claims* to goods do not simply "exist" but can only be ascertained discursively, which class for procedures of justification which must in turn be defined in normative terms as a matter of justice'; and
- 'leave the question of *injustice* largely out of the account – [and for example] equates someone who … is deprived of goods and resources as a result of a natural catastrophe with someone who suffers the same deprivation as a result of economic or political exploitation'. (Forst, 2017, p 122, emphasis in original)

A proper theory of justice instead 'must aim at *intersubjective relations and structures*, not *subjective* or *putatively objective states* of the provision of goods' (Forst, 2017, p 122, emphasis in original), and in particular must concern itself with the relations and structures of production.

Forst (2017, p 123) further argues that the opposite of justice is *arbitrariness* which in turn is at the core of domination. 'The basic impulse that opposes injustice in not primarily of wanting something, or more of something, but of no longer wanting to be dominated, harassed, or overruled as someone who has a claim and a basic *right to justification*' (emphasis in original). In this view, the central

> *political* essence of justice ... is who determines what is received by whom. On this conception the demand for justice is an emancipatory one. ... The person who *lacks* certain goods should not be regarded as the primary victim of injustice, but instead [the primary victim is] the individual who does not *count* when producing and allocating goods. (Forst, 2017, p 123, emphasis in original)

Forst links his conception of justice tightly to Marx's theories of exploitation and alienation (both of which are vital to making some people not *count*) and argues that it is in Marx's conception of the fetishism of commodities that the heart of the matter can be glimpsed.

Through commodity fetishism, relations between people appear as relations between things and the possibility of the free association of people (the precondition for justice as the right to justification) becomes impossible. Humans come to be dominated by 'an alien power' – the estranged commodities they make and their owners. By contrast: 'Freedom ... can only consist in this, that socialized man, the associated producers, govern the human metabolism with nature in a rational way, bringing it under their collective control instead of being dominated by it as a blind power' (Marx, *Capital*, vol. 3, in Forst, 2017, p 128). Like Marx, Forst (2017) holds that justice can only be approached through a radical transformation of the social relations of production, which in turn also requires a careful analysis of *injustice* as rooted in existing relations of production and distribution. Even more, Forst argues in a direct critique of Habermas, Rawls, and the Capabilities Approach, but in line with Kant, that *dignity* is a central object of justice, and dignity 'is violated when individuals are regarded as mere objects of social relations or *primarily as recipients of goods*' (Forst, 2017, p 129, emphasis added). Distribution is an insufficient basis for justice.

Standing somewhere between Habermas and Forst, Axel Honneth (1995) also holds that (re)distributive theories of justice are inadequate. For him, the core of injustice is non or misrecognition; justice requires recognition and respect. In turn, these must be founded on what Honneth (2004, p 355) sees as the 'three principles of recognition' – 'love, equality, and merit', which, slightly reformulated ('love ... equal treatment in law and ... social esteem') are also the 'three principles of social justice' (Honneth, 2004, p 358). Distributional theories of justice ignore this aspect of justice at their peril. For Honneth (2007), mis and nonrecognition are closely related to *reification*, which, in his hands, is transformed from a structural process emanating from within capitalism and its divisions of labour and necessarily fetishising social relations (Lukács, 1971 [1922]) to a kind of social psychology defined by 'intersubjective' power relations, which has the unfortunate effect of reducing the struggle for justice to a kind of demand for therapy. Despite Honneth's (2004, pp 362–363) rather arbitrary suggestion that claims for recognition must be assessed through a 'criterion of progress' and only those meeting such a criterion should be recognised, there is little radical in Honneth's arguments. As he himself says, his theory is highly affirmative of contemporary liberal capitalism.

In defence of Marxist theories of justice

While the question of justice was a central concern of the Frankfurt School, simultaneous developments within other branches of western Marxism in the 20th century also led to a good deal of scepticism concerning the validity of 'justice' as a Marxian concept. Perhaps most prominent in this regard was Louis Althusser's radical antihumanism and his promotion of a 'scientific' Marxism that sought to strip Marxian analysis of its normative dimensions and make it fully a *science* of society. Within Marxist philosophy more generally, there was a vigorous debate in the 1970s and 1980s about the status of justice within Marx's own thinking (see Geras [1985] for a review), with many arguing that:

- to the degree Marx traded in ideas of justice, he understood justice to be completely and fully determined by the stage of development of the mode of production (as Marx argued, exploitation in capitalism was not unjust, though slavery was); or
- it was at best a 'reformist' concept that had little room for revolutionary thinking and practice; or
- communism would be 'beyond justice' in that it would not be a scarcity-based mode of production and thus questions of distribution would not be questions of justice.

In these arguments, 'justice' for Marx was not transcendent, not universal, and not at all an ideal.

Defending the notion that Marx *did* hold a normative concept of justice (and condemned capitalism in terms of it), Norman Geras (1985; 1992), representing a significant number of other philosophers, argued instead that:

- For Marx, the argument that exploitation in capitalism was 'just' held only in the realm of exchange, where equivalents were traded (x amount of labour power for y amount of money representing its real value). Once one entered the 'hidden abode of production' where exploitation occurred, then any 'semblance' of justice rapidly disappeared and the unjustness of exploitation was rapidly exposed.

- Capitalism is based in theft. Marx saw the expropriation of surplus value as a kind of theft (and thus a question of justice) and he saw capitalism as having been born primarily through acts of theft, the thefts of enclosure, dispossession and colonisation.
- For Marx, 'standards of right' are sociologically grounded, which means they are 'constrained by the economic structure and resources of the given society' *not* that the standards for 'evaluating or assessing society must necessarily also be constrained by the same economic configuration' (Geras, 1985, pp 58–59).
- Demands for justice are not reformist but 'a relatively independent contribution to processes constituting the human agency of revolutionary change [and] the formation of a desire and consciousness for socialism' (Geras, 1985, p 60). Marx's sense of justice was in this sense both juridical and normative.
- The 'needs principle' ('from each according to their abilities') which is at the root of Marx's theory of justice concerns modes and relations of production but is also distributive in that Marx recognised that distribution according to need is fairer than distribution according to ability, merit or ownership.

Taken as a whole, this defence of Marx as a justice theorist argues that in his dialectical analysis, Marx held justice to be always *actually-existing* (and thus limited and ideological) and a *normative ideal* (and thus, as Engels put it, the stick against which these actually-existing conditions could be measured). The core of justice, for Marx and in terms of this defence, was rooted in the 'needs principle' and could be described as a kind of productive–distributive theory that required a thoroughgoing transformation and reorganisation of the relations of production so that a more just distribution (based in need) could be achieved.

Radical justice and the social sciences

Despite the efforts of the Frankfurt School, development of radical theories of justice within the various social sciences was relatively muted during the middle decades of the 20th century and after 1971 the overwhelming influence of Rawls turned many radical thinkers away from direct engagement with justice theory. One major, early exception was in the field of geography, where the work of David Harvey has been of inestimable importance. In the wake of Geras' (and others') defence of explicitly Marxist theories of justice as well as the collapse of state-socialist Eastern Europe and the Union of Soviet Socialist Republics (USSR), Marxian and other socialist forms of justice theorising have enjoyed something of a renaissance, as scholars (and activists) have come to better recognise the analytical potential of a dual notion of justice that simultaneously assesses 'actually-existing' conditions of justice and drives towards normative ideas.

Marxism and the geography of justice

Having moved to Baltimore in the wake of the nationwide urban unrest in the United States in 1967–1968, the geographer David Harvey quickly grew disenchanted with his position as the discipline's foremost philosopher of positivist epistemology. For all the power of its statistic and other mathematical tools, positivist geography had no way to explain, much less assess, either the conditions leading to or the eventual results of this unrest. Harvey turned instead to questions of justice. Strongly influenced at first by Rawls' (1971) recently published *Theory of Justice*, Harvey (2009 [1973]) sought to 'spatialise' liberal theories by asking what a just *spatial* distribution in the (American) city would look like. It was not long, however, before he was dissatisfied with this line of inquiry too, since, he discovered, questions of *equity* (central to liberal theories of justice as codified by Rawls) were inevitably reduced to questions of *efficiency* and thus the promotion of a kind of technocratic reasoning that left little room for understanding either the sources of injustice (and thus served to perpetuate them) or the real interests, desires and needs of people who live in cities.

Harvey (2009 [1973]) therefore turned to what he called 'socialist' formulations of the justice question and launched a still ongoing effort to both retheorise Marx spatially and develop a theory of justice adequate to Marxism's revolutionary aims – with Marx's own insights as vital fuel for his arguments. His work is impossible to briefly summarise, but three central principles of his work can be identified:

1. Any approach to theorising justice must be *dialectical* (in the sense developed by Ollman, 1991), seeking to understand the totality of society and its relations.
2. It must also be attentive to the relationship between the universal and the particular and thus the *geography* of difference.
3. All justice theory must thus incorporate efforts to understand what would constitute a *just production* of geographical difference (Harvey, 1996).

Though seeking radical approaches to justice, Harvey in some ways never really leaves the Rawlsian fold: he too is concerned with theorising a just 'difference principle'. The primary difference between the two is that for Harvey, any just difference principle must be rooted in the production of geographical space as a fundament of the mode of production rather than in distribution. In this sense, Harvey's position largely accords with Geras', but with the significant addition of a strong spatial component, which is to say that for geographers like Harvey, any Marxian and radical theory of justice has to take as a starting point both the geographical preconditions for, and the geographical results of, the dialectical interplay of universal and particularistic forces. As in feminist theories of global justice (see Chapter 4), in Harvey's theory lies a central concern with how seemingly just interventions in one place (or at one geographical scale) may have decidedly unjust outcomes in another.

Geographers have developed Harvey's arguments in three general directions – towards theories of spatial, landscape and environmental justice, and the results of these inquiries are detailed in the chapters covering these topics. In all three areas the influences of Marxian and radical approaches to justice are apparent. But one key concept should be highlighted here. For Harvey (2012), all his sophisticated work on spatialising radical philosophies of justice has led to a central concern with how peoples' desire for justice often resolves into a 'cry and demand' for the 'right to the city'. The language of cry and demand, as well as the original concept of a right to the city derive from the French philosopher/social theorist, Henri Lefebvre (1996 [1967]), who, writing on the centenary of Marx's *Capital*, argued that through an effective right to the city, which entails an effectively just city (and countryside), all people would have a right to:

- Centrality – access to the heart of the city, but also and especially access to the power to determine the city's development and use.
- The oeuvre – the ability to be centrally involved in the making of the city as an ongoing work, not a once-and-for-all product.

Geographers, urban sociologists, radically inclined anthropologists, architects and other spatial thinkers have latched onto this idea, and much research thus concerns itself with the question of what the conditions of possibility are – or could be – for the achievement of a right to the city, understood both as a congeries of spaces and places and as a metonym for the right to social life, to *species being*.

A focus on injustice and the promises of radical justice for social science research

Radical approaches to justice have been particularly important in encouraging a strong focus on – making a priority of – *injustice* (Barnett, 2017). A radical core has been retained in this work when it has focused on the roots of injustice in the social relations of production rather than procedure or distribution. This does not imply that questions of distribution are absent from radical theorising in the social sciences, but rather that such questions are subordinate or secondary to questions related to production. By moving from injustice (which is what 'actually-existing' justice looks like on the ground) to questions of just modes and relations of production – coupled with people's fundamental right to justification and right to the city (as a metonym for society as well as a spatial reality) – radical theories of justice have profound implications for the social sciences. They fully reorient how we conceive of the project of justice theorising and especially the struggle for justice. They require social scientists to understand:

- The *subject* of justice as 'people' in general (not sovereign individuals), who, in their capacity for rational deliberation and their emancipatory potential seek the freedom (in solidarity) to develop their *species being*; in actually-existing society, this practically means the oppressed and exploited classes – precisely those who most need the right to justification.
- The *object* of justice as the mode of production and its social relations; in actually-existing society, this means its processes of exploitation, alienation and fetishisation, each of which needs to be transcended to create a just society founded in mutuality.
- The *domain* of as justice being scalarly complex while the universal and particular are mutually determinative; in actually-existing society, this requires the development of both deliberative and confederationist modes of governance.
- The *social circumstances* of justice as occurring at the points of production – both the production of goods (and thus needs fulfilment) and the production of the historical–geographical preconditions within which this more narrow form of production unfolds; in actually-existing society this means justice occurs both in the struggle for 'industrial democracy', the 'right to the city' and 'right to landscape' (see Chapter 13), as well as in the creation of forums for the 'right to justification'.
- The *principles* of justice as entailing a radical reconstruction of deliberation (to address imbalances of power), the construction of mutuality, and the centrality of solidarity; in actually-existing society this requires, for example, the abolishment of 'consultation' and its replacement with active deliberation and appropriate mechanisms to make this possible.

Such then, is the challenge that radical approaches to justice – as outlined in these five aspects of justice – pose to social scientists: How do we understand actually-existing injustice? And how to we get from that understanding to something closer to Engels' 'stick' – the radical ideal of justice as human emancipation? If not in so many words, that has been precisely the heart of the question insistently posed (however incompletely or contradictorily answered) from Fourier to Forst and beyond.

References

Adorno, T. (1972 [1954]) Ideology. In *Aspects of Sociology*. Boston, MA: Beacon Press, pp 182–205.

Adorno, T. (1981) *Prisms*. Cambridge, MA: MIT Press.

Barnett, C. (2017) *The Priority of Injustice*. Athens, GA: University of Georgia Press.

Forst, R. (2004) The limits to toleration. *Constellations*, 11(3), 312–325.

Forst, R. (2010) The justification of justice: Rawls and Habermas in dialogue. In J.G. Findlayson and F. Frayenhagen (eds), *Habermas and Rawls: Disputing the Political*. New York: Routledge, pp 153–180.

Forst, R. (2012) *The Right to Justification: Elements of a Constructivist Theory of Justice*. New York: Columbia University Press.

Forst, R. (2014) *Justification and Critique: Towards a Critical Theory of Politics*. Cambridge, UK: Polity.

Forst, R. (2017) *Normativity and Power: Analyzing Social Orders of Justification*. Oxford, UK: Oxford University Press.

Geras, N. (1985) The controversy about Marx and justice. *New Left Review*, 150, 47–85.

Geras, N. (1992) Bringing Marx to justice. *New Left Review*, 192, 37–69.

Habermas, J. (1962) *The Structural Transformation of the Public Sphere*. Cambridge, MA: MIT Press.

Habermas, J. (1981) *Theory of Communicative Action*. Boston, MA: Beacon Press.

Habermas, J. (1990) Justice and solidarity: On the discussion concerning stage 6. In T. Wren (ed), *The Moral Domain: Essays in the Ongoing Discussion between Philosophy and the Social Sciences*. Cambridge, MA: MIT Press, pp 224–251.

Habermas, J. (1992) *Between Facts and Norms*. Cambridge, MA: MIT Press.

Habermas, J. (2001) *The Postnational Condition*. Cambridge, MA: MIT Press.

Habermas, J. (2012) *The Crisis of the European Union*. London: Wiley.

Habermas, J. (2014) Plea for a constitutionalization of international law. *Philosophy and Social Criticism*, 40(1), 5–12.

Harvey, D. (1996) *Justice, Nature, and the Geography of Difference*. Oxford, UK: Blackwell.

Harvey, D. (2000) *Spaces of Hope*. Edinburgh: Edinburgh University Press.

Harvey, D. (2009 [1973]) *Social Justice and the City*. Athens, GA: University of Georgia Press.

Harvey, D. (2012) *Rebel Cities*. London: Verso.

Hayden, D. (1976) *Seven American Utopias: The Architecture of Communitarian Socialism, 1790–1975*. Cambridge, MA: MIT Press.

Held, D. (1980) *Introduction to Critical Theory: Horkheimer to Habermas*. Berkeley, CA: University of California Press.

Hobsbawm, E. (1962) *The Age of Revolution*. London: Weidenfeld & Nicolson.

Honneth, A. (1995) *The Struggle for Recognition: The Moral Grammar of Social Conflicts*. Cambridge: Polity.

Honneth, A. (2004) Recognition and justice: Outline of a plural theory of justice. *Acta Sociologica*, 47(4), 351–364.

Honneth, A. (2007) *Reification: A Recognition-Theoretical View*. Oxford: Oxford University Press.

Horkheimer, M. (1972) *Critical Theory: Selected Essays*. New York: Continuum.

Kropotkin, P. (1902) *Mutual Aid: A Factor of Evolution*. https://theanarchist library.org/library/petr-kropotkin-mutual-aid-a-factor-of-evolution

Kropotkin, P. (2006 [1892]) *The Conquest of Bread*. Oakland, CA: AK Press.

Lefebvre, H. (1996 [1967]) The right to the city. In *Writings on Cities*. Oxford, UK: Blackwell, pp 63–181.

Lukács, G. (1971 [1922]) *History and Class Consciousness*. Cambridge, MA: MIT Press.

Marcuse, H. (1955) *Eros and Civilization*. Boston, MA: Beacon Press.

Marx, K. (1844) *On the Jewish Question*. https://www.marxists.org/archive/ marx/works/1844/jewish-question/

Marx, K. (1932 [1844]) *The Economic and Philosophic Manuscripts*. Moscow: Progress Publishers.

Marx, K. (1932 [1846]) *The German Ideology*. Moscow: Progress Publishers.

Marx, K. (1976 [1856]) *Grundrisse*. London: Penguin.

Marx, K. (1987 [1867]) *Capital*. New York: International Publishers

Merrifield, A. and Swyngedouw, E. (eds) (1995) *The Urbanization of Injustice*. New York: New York University Press.

Ollman, B. (1991) *Dialectical Investigations*. Oxford, UK: Blackwell.

Pierce, A. (2017) Justice without solidarity? Collective identity and the fate of the 'ethical' in Habermas' recent political theory. *European Journal of Philosophy*, 26, 546–568.

Pettit, P. (1982) Habermas on truth and justice. *Royal Institute of Philosophy Lectures*, 14, 207–222.

Proudhon, P.J. (1923 [1851]) *General Idea of the Revolution in the Nineteenth Century*. London: Freedom Press.

Rawls, J. (1971) *A Theory of Justice*. Cambridge, MA: Harvard University Press.

Schick, K. (2009) 'To lend a voice to suffering is a condition for all truth': Adorno and international political thought. *Journal of International Political Theory*, 5(2), 138–160.

Wieck, D. (1978) Anarchist justice. *Nomos*, 19, 215–236.

Wolff, J. (2017) Karl Marx. In E.N. Zalta (ed), *The Stanford Encyclopedia of Philosophy*. https://plato.stanford.edu/archives/win2017/entries/marx/

Young, I.M. (1990) *Justice and the Politics of Difference*. Princeton, NJ: Princeton University Press.

6

Radical Justice Through Injustice: Postcolonial Approaches

Johanna Ohlsson and Don Mitchell

Introduction

Postcolonial theories in general, and concerning justice and injustice more particularly, present a significant challenge to approaches to justice offered in most of the preceding chapters. They directly challenge the dominance of 'western', 'Global North' or 'Eurocentric' thinking that mark these other traditions or schools (and which were complicit in European imperialism and colonialism), even if they remain significantly indebted to (some of) the epistemologies and problematics that define the western canon. Postcolonial theory is thus not a disciplinary field, as such, but has arisen within a range of fields (and activist formations) as a movement aiming towards the transformation of both scholarly and political practice.[1] As a theoretical orientation, postcolonialism can be described as a way of viewing and understanding the world that cuts across political, economic, cultural, symbolic and linguistic spheres to address the structures of power as well as knowledge production, while asserting the *centrality* of peoples, societies and countries in the (historical) *periphery*. That is to say, postcolonialism aims at reconstructing western intellectual formations and norms while seeking to turn such relations of power upside down in order to refashion the world from below (Young, 2012, p 20). At its core, the postcolonial tradition is a critical approach to knowledge concerned with unjust, unequal and asymmetrical relations of power, and with domination, oppression and the workings of western hegemony more generally.

Overview of central ideas and scholars

One of the main contributions of postcolonial thinking has been to promote the voices of the oppressed, often from the Global South, and thereby create fuller awareness of the asymmetrical power relations governing interactions of the oppressed and the oppressors.[2] This orientation towards power leads to the predominant way *justice* is understood within postcolonialism: *justice as non-domination*. Similar to the understanding of justice developed by some theorists of the feminist approaches to justice (Chapter 4), within Marxism and critical social theory (Chapter 5) and within the Capabilities Approach (Chapter 8), postcolonial approaches to justice focuses on *injustice* and on the conditions of possibility for being *free from domination*. Within this overall orientation, postcolonial theorising has sought to understand the specific ideological processes by which the west defines itself against its other ('the rest') (Said, 1978; 1993; Bhabha, 2004 [1994]); the European colonial legacy in the Global South (Fanon, 1967 [1952]; 1963 [1961], and the work he inspired); and the more general workings of oppression in its many forms (Spivak, 1999 [1988]).

Postcolonialism is of crucial importance in relation to theorising justice because it seeks to create means – within and beyond scholarship – for peoples previously and currently oppressed to speak for themselves. It offers important theoretical tools for making visible asymmetries of power, both within and between groups, societies and states. In doing so it contributes to awareness of the privileges as well as negative stereotypes generated by racism, sexism and other forms of discrimination.[3] In particular, it forces critical considerations of how racism, sexism, western domination (imperialism, colonialism) and other forms of oppression might be discursively built into dominant, western theories of justice, even of the most radical sort. Postcolonial theory, in other words, highlights the intentional and unintentional exclusion and silencing of 'the other' within justice theory. It interrogates the constitutive 'absences' of western theorising.

By including the voices of women, people of colour, people from the South, people from different socioeconomic backgrounds, and so forth, the implicit – and sometimes explicit – masculinist, White, colonialist and elite foundations of western justice theorising are highlighted, bringing to light the production of past, present and future injustices.[4] Postcolonial theory starts from the recognition that the legacies of the transatlantic slave trade, the colonisation of the Americas, Africa and parts of Asia, and the subsequent era of imperialism still ramify through the everyday lives of subjugated peoples. Postcolonial theories of (in)justice, therefore, are rooted in struggles for liberation, struggles that range from freedom from slavery to the decolonisation of knowledge and knowledge production (see, for example, Mohanty, 2003). Most commonly, though, postcolonialism has developed as a response to the need to understand, and confront, colonial exploitation and to support the liberation movements that continue to struggle against imperialist and colonial legacies.

Given the range of histories (geopolitical as well as scholarly) coalescing under the umbrella of postcolonialism, the discussion that follows will necessarily be selective. But what unites all postcolonial thinking – and thus its implications for theories of justice – is a focus on the unjust and unequal structures (of politics, economy, knowledge, and so on) that result from the history of colonialism and imperialism.

Key ideas: debates and critiques

In many ways, it is difficult to identify a *particular* postcolonial theory of justice. While postcolonial theories frequently invoke 'justice' or 'social justice' as normative values, and postcolonial scholarship is just as frequently dedicated to revealing and addressing injustice, little work has been addressed to theorising justice *as such*.[5] This might very well be a deliberate choice and a way of distancing postcolonial thinking from the dominant discourse on justice. Instead, what postcolonial theory and scholarship offers are a set of ontological and epistemological presuppositions that challenge mainstream theories of justice while providing a foundation for their reconstruction along postcolonial, or decolonised, lines. In particular, by seeking to 'decentre' the west as the yardstick against which all knowledge (and justice) is measured, postcolonial theory might offer means towards something like the promise of cosmopolitan justice outlined in Chapter 3, by seeking to decentre nation states as the (primary) object of justice, while also critiquing and rejecting the weight of its colonising western and European origins. This section of the chapter seeks to outline some of the ways in which these ontological and epistemological presuppositions have been – or can be – understood to shape justice theorising, especially in relation to the questions of the *subject* and *object* of justice. Matters related to the domain, circumstances and principles of justice will be raised in the following section.

The subject of postcolonial justice

Postcolonial scholars challenge both liberal accounts of historical development and change (which underestimate the relations of power structuring 'the other' *as* other, and thus excluded from the full benefits of liberal sovereignty and citizenship – see Chapter 1) and the Marxist primacy of *class* as the driving force of historical change (see Chapter 5). They argue instead for a more intersectional understanding of identity (often in line with feminist approaches, see Chapter 4), within which *race* and *subalternity* are particularly salient categories. The dynamics of racism and its associated hierarchies are understood to be fundamental forces shaping historical development and change (Nair, 2017). Within postcolonial scholarship, then, the *subject* of justice is neither the liberal individual nor the working class, but the subaltern and the oppressed.

Postcolonialism thus challenges and seeks to upset the privileged subject of justice in both liberalism and Marxism which is imagined to be White, western and male.[6] It insistently asks: *Who* is being listened to or heard (and who is being silenced)? *Whose* experiences are acknowledged? *Who* is an expert and seen as capable of producing knowledge? The systematic exclusion of some subjects from being the answer to these questions is what Miranda Fricker (2007; 2013; 2018) calls *epistemic injustice*.[7] Epistemic injustice is systematic silencing, misinterpretation or misrepresentation that diminishes or destroys one's standing as a subject of justice. For Fricker, epistemic injustice entails both *testimonial injustice* (having one's words not trusted, typically on grounds of race, gender or other markers of identity and social position) and *hermeneutic injustice* (having one's experiences systematically misinterpreted because there does not yet exist sufficient language for interpreting them, largely because of the systemic exclusion of certain subjects as producers of 'valid' knowledge). If not always developed explicitly in these terms, postcolonial theory is centrally concerned with highlighting and seeking to rectify epistemic injustice and establish (post)colonial subjects as full subjects of history – and justice.

Prominent in this project of asserting the subaltern as a subject of justice has been Gayatri Spivak, whose work focuses explicitly on understanding structures of domination.[8] In particular, she has considerably developed the Gramscian notion of 'the subaltern'.[9] In her (frequently rewritten and republished) essay, 'Can the subaltern speak' (1999 [1988]),[10] Spivak transforms the analysis of colonialism through an uncompromising argument that both affirmed the contemporary relevance of Marxism and sought to transcend it through the development of deconstructionist analyses. She argued that whatever the efforts of the subaltern subject, the reality of the history of colonial oppression means that this subject does not get heard. The question Spivak raises is precisely one of the *force* of epistemic injustice: given the history and ongoing processes of oppression *can* the voice of the subaltern ever really be heard? Is hermeneutic injustice a necessary and structuring part of colonial relations and thus the postcolonial condition? One conclusion that could easily be drawn from Spivak's insistence that the subaltern *cannot* be heard is that human rights discourse and practice, based as they are on an assumption of liberal, rational subjectivity, will be unavailing, given the structural forces that exclude many from just this subjectivity. As Drucilla Cornell (2010, pp 110–114) has argued, any legitimate human rights discourse requires *recognition* of new subjects of justice – the excluded and oppressed subalterns themselves.

As Frantz Fanon (1967 [1952]) long ago recognised, the problem confronting the subaltern subject was not so much misrecognition, as a form of recognition that insisted on the subaltern's inferiority, and which the subaltern nearly inevitably internalised. The subject of justice in this sense is to a large degree the *racialised* subject, but it is also the subject that learns to question its racialised *subjectification*. To some degree, Fanon's arguments are compatible with aspects of liberalism (see Chapter 1) in that he starts from the assumption of each human being as free and equal (and from the assumption that a just society is one that recognises and protects this fact), yet departs from it in his insistence on understanding the reality of individuals' lives as thoroughly structured through the hierarchical relationship between the coloniser and colonised. The colonised are fully subjected to and permanently marked by the forces that dominate them (Fanon, 1967 [1952]).[11]

The object of postcolonial justice

The *object* of justice in postcolonial theory is therefore the historical and contemporary *relations and institutions of power* (including especially geopolitical power) that oppress and dominate (post)colonised peoples and produce epistemic as well as other types of injustice. This object of justice is understood in specifically geographical terms: the question of where in the world dominating power, including the power to produce knowledge, emanates. But it is also understood in analytical terms: the question of where (and on what) the focus of analytical attention is centred. Besides institutions and relations, as such, such analytical attention is equally focused on the norms and values on which these institutions and relations are founded.

Commonly, both in justice theory (as is made clear in the preceding chapters) and academia at large, the yardstick, or point of reference, is the west, not only because it is where the centres of intellectual power reside and where most (even postcolonial) theorists are based, but also because *it* is understood to be the pivot of history. In this sense, the object of justice in postcolonial thinking can be more pointedly understood as the *distribution of recognition*, and the political goal of postcolonial theory is to radically transform that distribution.[12] The goal is not only to 'open up space' for theorising from Asia, Africa and Latin America but to radically transform the conditions of knowledge production by *centring* voices from these regions (Hegde and Shome, 2002; Robinson, 2005; Mignolo and Walsh, 2018).

Such a transformation is vital because, as Edward Said (1978) carefully detailed in his explication of the centrality of 'Orientalism' in western philosophy, culture and geopolitical practice are definitive. 'Western knowledge of the Eastern world', Said showed, usually depicted the latter as irrational and its denizens as weak 'others' (though which Europeans confirmed their own superiority). On the one hand, Said's analysis makes doubly important the 'who' question concerning the subject of justice discussed previously: can justice theory make room for a non-western 'other' (and if so, how)? On the other hand, it forces attention on the object of justice – the relations and institutions of power – in particular, by showing how 'Orientalism' was itself one of these relations and institutions of power, and a deeply seated one, at that. For Said, 'Orientalism' is at once:

- an academic tradition or field, indispensable to determining who counts as a rational subject and thus entitled to the just fruits of the Enlightenment;
- a worldview or 'style of thought based upon an ontological and epistemological distinction made between "the Orient" and (most of the time) "the Occident"'; and
- a powerful political instrument of domination.

Together these help to form what Gramsci (1971) would have called a 'common sense', a sense of taken-for-granted presuppositions that can only be undone (and replaced instead with 'good sense') through concerted political struggle.

But 'Orientalism' and other forms and impositions of power are never fully complete or unassailable. Building on Said's work, Homi Bhabha (2004 [1994]) has focused in particular on the inherent vulnerabilities within the colonial discourse and thus power (see Ashcroft et al, 2007, p 37), revealing how they operate through the production of hybrid, rather than unitary, subjects and how mimicry (of the dominant, by the dominated) can be an important source of resistance to dominating power.[13] Such mimicry (among other processes) introduces an important ambivalence into the colonial relation exacerbating its vulnerability.[14] Though not directly addressing theories of justice, Bhabha's arguments – which are foundational for contemporary postcolonial theory – are important to such theories precisely because they show that the object of justice in the postcolonial age (the relations and institutions of power) is itself vulnerable to 'hybridisation', which is to say its remaking.

The object of justice, therefore, is that set of relations and institutions that *centre* power (or rather, as Michel Foucault [1980] termed it, *power-knowledge*) in the west, both historically and now, and which thereby pervert the *distribution of recognition* so as to *systematically* devalue, exclude, subjugate and dominate the racialised, gendered, colonised *other*, which is to say that the object of justice for postcolonial thinking is those relations and institutions that promote *epistemic injustice*.

From substance, procedure and distribution to rectificatory justice

Postcolonial theory has been particularly concerned with questions of the diverse ways in which postcolonial subjectivities are constituted and resisted through discursive practice (Kohn and Reddy, 2017). Questions of representation are thus central and help account for the predominance of literary theory in the field, despite its genuine multidiciplinarity (Lazarus, 2004). Such questions of representation tend in turn to pivot on the relations between self and other and the injustices that arise in these relations. For this reason, explicit postcolonial theorising of justice has tended to focus more on questions of *substance* than *procedure*.

Substance, procedure and distribution

Such a focus on substantive justice has entailed a (largely implicit) reconsideration of the *domain* of justice. In *Postcolonial Justice*, the most prominent of the small body of work explicitly theorising justice through a postcolonial lens, Anke Bartels et al (2017) suggest that any truly postcolonial justice cannot be merely local, based only on local knowledge and experience. Rather, what is needed is 'a genuinely postcolonial reformulation of planetary justice' that 'must also learn to de-privilege western re/visions of ethics and justice, be it in the form of Derrida's radicalisation, of Kant's cosmopolitan hospitality, or Levinas' riffing on Heidegger' (Bartels et al, 2017, p 5). Like Dipesh Chakrabarty (2012) and others, Bartels et al argue that the concepts developed by Kant, Heidegger, Derrida and Levinas which recapitulate and reinforce colonial legacies and rely on them will reinforce such legacies rather than decolonise knowledge and power. Such ideas are inseparable from their colonial entanglements and racist legacies (Chakrabarty, 2012).

As substantive injustices are identified, therefore, solutions must be found that are untainted by these entanglements and legacies. It is the great failing of postcolonial thinking, however, to not offer much by way of a hint as to what these might be beyond Spivak's (2012) desire for 'planetarity' (which Bartels et al [2017] echo). Precisely what Spivak means by this is hard to know. One scholar who has tried to interpret Spivak's theory of planetarity, however, is the postcolonial geographer Joel Wainwright, who interprets it as follows:

> Let us no longer speak of globalization, the global scale, and the like; instead let us think of ourselves as living on a planet. ... 'The planet,' [Spivak] explains, 'is in the species of alterity, belonging to another system' (Spivak 2012, p 338), one beyond our control and even representation. (Wainwright, 2013, p 70)

In his reading of Spivak, Wainwright (2013, p 70) emphasises that her understanding of the world *qua* 'planet' is different from what he argues is the more common conception of the world *qua* 'globe'. He clarifies that in her view, 'the planet is one of those things that can never be a thing, but a thing-in-itself, something that we know is there, though we can never directly grasp as an object with our senses'.[15] Wainwright (2013, p 73) interprets Spivak as seeing the task at hand being to understand places 'not geographically, or through its ally, area studies, but as a debate, not as an object that exists empirically but as a text, or a group of texts' (quoting Ismael, 2005). He questions this abstraction, asking 'if not as empiricists, [how] are we to think planetarity?', highlighting a tension in abstract theorisation and empirical work, which is not only a feature in postcolonial thinking on justice and injustice but commonly seen also in other scholarly traditions.

Not all postcolonial theorising in relation to justice is this idealist, however. When postcolonial theorists have turned to more specific concerns with *geographical* distribution some of postcolonialism's radical potential is regained. In line with feminist theorists like Iris Marion Young (2000; 2011), David Turnbull (2017) argues that the maldistribution of wealth, power and knowledge has come about under present systems of law and justice – considered, through western and liberal eyes, to be fundamentally procedurally just, even if sometimes unjust in their actual operation and outcomes – and so it is these very systems that must be resisted. Analytically, therefore, he advocates adopting a spatial performative approach to counter the injustices that have arisen through these seemingly just institutions. By spatial performative, Turnbull (2017) seems to mean paying close attention to the spatial practices by which powerful institutions, rooted in historical colonial and imperial centres – Europe, North America, and so on – impose their will in peripheral places. He focuses particularly on the spatial performance of law that allows for 'ongoing processes of enclosure and dispossession that not only displace indigenous peoples and destroy their cultures but also destructively appropriate the environmental and knowledge commons and drive the flourishing inequity' (2017, p 5). As he (2017, pp 7–8) argues, '[i]n practice, we make our world in the process of moving through and knowing it, and the making of knowledge is simultaneously the making of space. But this performative coproductive process is itself performatively displaced and erased in the technologies of representation, maps, legal texts, and codes'. Turnbull's spatial performative approach offers one answer to how some of the *principles* – or 'how' – of justice are conceptualised in postcolonialism.

A second, yet somewhat related, answer could be seen as grounded in a movement in international law, *Third World Approaches to International Law* (TWAIL). Scholars within TWAIL make their critiques principally on an institutional level, often arguing that the 'regime of international law is illegitimate' as it 'legitimizes, reproduces and sustains the plunder and subordination of the Third World by the West', as claimed by Makau Mutua and Antony Anghie (2000, p 31). Building on Arif Dirlik (1994), they see the terms postcolonial and postcoloniality as referring to 'an intellectual trend in many western universities toward reclaiming Third World concerns within the general framework of postmodernism' (Mutua and Anghie, 2000, p 32). This might be seen as closer to ideal theorising than geographically oriented empirical research, but it still offers important insights on approaches to (perhaps primarily – but not only – legal) justice within postcolonial thinking. The approaches within TWAIL include 'critical, feminist, post-modern, Lat-Crit Theory (Latina and Latina Critical Theory Inc.), postcolonial theory, literary theory, modernist, Marxist, critical race theory' to name a few, showcasing the diversity of TWAIL not being only a postcolonial approach (Gathii, 2011). This also indicates the diversity in ongoing conversations within postcolonial thinking, as TWAIL scholars are both leveraging critiques towards some ideas in postmodern and postcolonial thinking, and furthering the state-of-the-art in the same.

Rectificatory justice

A third answer to how postcolonial thinking approaches the 'how' (and also the 'when') of justice is through the matter of rectification.

Indeed, in some ways, the most radical aspect of postcolonial thinking about justice is precisely this focus on *rectificatory justice*, which has been relatively neglected in more mainstream theories of justice (Collste, 2015; Roberts, 2017). Rectificatory justice is concerned with setting unjust settings right, with righting injustice. It is not the same as retributive or corrective justice, which is typically concerned with legal and judicial processes that aim to 'do justice' by punishing wrong acts.[16] Such corrective justice has rectificatory elements, but they are less prominent than in current attempts to develop theories of rectificatory justice suitable for understanding and addressing the harms of colonialism and imperialism. If procedural justice is about fair procedures and distributive justice is about fair distributions, rectificatory justice is about *righting* the unfair treatment of some by others and especially about addressing the practices and institutions that allow or promote such unfair treatment (Roberts, 2017, p 516). Rectificatory justice is commonly understood as implicating more than judicial procedures. In Rawlsian terms (to the degree that such terms are not ruled out by the postcolonial project) it could be understood as focused on the organisation and operation of the basic structure.

As such rectificatory justice is closely connected to questions of political and social responsibility (Young, 2011; Collste, 2015), which in turn implicate matters of *reparation, restoration, compensation* and *apology*. Each of these matters are central concerns of rectificatory justice: Of what is just reparation for past harms – for example, slavery – comprised? How can violations of bodily, cultural or environmental integrity be restored? To what degree can individuals and communities suffering harm be compensated so that such harm may continue (a crucial question in some theories of environmental justice, as we will see in Chapter 9)? Who, or what agencies, should apologise – and to whom? What constitutes a *just* apology?[17] Addressing such questions offers the further potential for understanding the mechanisms that instantiate injustice (no matter how 'just' they may appear in any historical-geographical context) and thus work towards a more just society.

Rectificatory justice is, in this sense, concerned with the *social circumstances* – the 'when' – of justice, and as such should not be conflated with retributive or even restorative justice (though seeking the latter might be a necessary consequence of rectificatory processes: restoration or reparations might be a *means* of righting past and present wrongs). The kind of punishment that is central to retributive practices of justice is not possible in the same way for rectificatory justice, since the wrongful acts commonly began generations ago (Collste, 2015), lasted for decades or centuries, and implicated the total membership of whole societies, with differently positioned members shouldering different levels of responsibility. This question of responsibility is therefore crucial (see, for example, Young, 2011). *Who* is to be held responsible – and for *what*? How can culpability be apportioned, especially over current generations for past harms? Can contemporary populations be held accountable for the colonising practices (from land theft, slavery and plundering to genocide) their ancestors implemented and succeeding generations supported? Within this, how can internal opposition to the colonising project be assessed (since it was rarely absent even within the most imperialistic of powers)? How can the responsibility of seemingly 'passive' beneficiaries of colonial privileges (like the industrial working classes of Europe) be apportioned? These seem to be central questions for postcolonial thinking on rectificatory justice and indicate the need for historical and contextual sensitivity.

One, if still partial, answer to these questions within postcolonial theorising has been to focus on the symbolic and substantive value of memory and apology (see, for instance, McGonegal, 2009), for example, assessing the importance of ceremonial return of ancestors' remains (and cultural artefacts) formerly held in museums in colonial centres (Collste, 2015; Bartels et al, 2017).

What makes the idea – or rather the circumstances – of rectificatory justice radical is precisely the fact that it seeks to hold current populations accountable for past wrongs. Yet radical, rectificatory justice is not a new idea. It was already articulated by Aristotle (1999), and its content and meaning have been debated ever since, though largely as a minority tradition. While rectificatory justice hardly figures at all in contemporary mainstream liberal thinking about justice, philosophers such as Hugo Grotius and John Locke considered rectificatory aspects of justice in their work (Roberts, 2011). And Kantian and deontological theories of justice generally have been foundational to the contemporary development of rectificatory theories in postcolonialism. Rather than seeing such Kantian (or Lockean) roots as irremediably contaminating (as they can be figured within some of the most extreme versions of postcolonial thought, which seeks to dismiss all contamination by 'the west') these roots can provide the opportunity for understanding how the historical eras within which these foundational philosophers worked still shape the global order, and thus the ongoing legacy of colonialism provides a strong case for fighting for global rectificatory justice (Collste, 2015). In sum, the 'when' of justice is in the past – as well as in the present's responsibility for the past.

This view of justice as a rectificatory process, however, is not universally shared among postcolonial thinkers. Frantz Fanon (1967 [1952]), for example, seems to have been aiming towards a theory of justice that was independent of history, in the sense that the future should not necessarily be bound to history and its determinations. If African nations fail to liberate themselves from history – as well as from their colonial masters – Fanon argued at the end of *Black Skin, White Masks* (1967 [1952], pp 229, 231), they will become a prisoner of that history, a subjectivity Fanon absolutely refused: 'I am not a prisoner of history ... I as a man of colour, to the extent that it becomes possible for me to exist absolutely, do not have the right to lock myself into a world of retroactive reparations'. Instead, Fanon argued for a radically existential and liberationist stance towards justice: as with much Marxist and anarchist theorising, justice consists in human liberation (see Chapter 5). In a deeply fascinating, but sadly incomplete essay written at the end of her life, Iris Marion Young (2011, ch 7) argues that Fanon's stance requires that 'we neither seek guilt for the past in the present nor try to forget it' (p 172) and advocates instead her 'social responsibility model' as a forward-looking means to apportion responsibility, rather than guilt or liability. This does not necessarily preclude reparations for past harm, but any such reparations must be assessed in relation to possibilities for transforming the present and shaping a more just future. The 'when' of justice is in the future, the seeds of which are currently being laid.[18]

Postcolonial thinking on justice and injustice in social science research

The primary relevance of postcolonial approaches when thinking about justice is that, as with Marxian and feminist approaches, it requires a critical rethinking of our conceptions of justice – their scope (subject, object, domain, social circumstances, principles) as well as their silences, absences and possible complicity in social structures of domination and oppression (see Mignolo and Walsh, 2018). Radical approaches to justice adopt a critical lens to ask – to require an assessment of – who is harmed, and who benefits from the current arrangement of institutions, social relations, economic practices and exercises of power – and how.

Like other radical theories of justice, postcolonialism also often denies the radical individualism of liberalism (even if, as with Fanon, it sometimes promotes the importance of individuals' existential *liberation*). As such it requires a focus on the relations and institutions of power (the primary *object* of justice in postcolonialism), as they shape collective life, the exploitation of resources, the use and abuse of the environment, and so forth. It requires a quest for understanding how 'others' are formed, and who they are, how they are being heard and listened to. Recent postcolonial-inspired work has sought, therefore, to extend the concept of 'the other' to include nature (see, for instance, Plumwood, 2003).

As has been shown in this chapter, postcolonial thinking on and in relation to justice is a scholarly but sometimes also an activist endeavour. Some of the social movements most clearly connected to these lines of thought are perhaps the *Sumak Kawsay*, *Suma Qamaña* or, in its Spanish version, *Buen Vivir*. This is often conceptualised as an alternative philosophy and view of life corresponding to various Indigenous approaches, but also as a political platform for different visions of alternatives to the dominating discourse on development as well as non-western thoughts on the social aspects of life. It has also made its way into governance structures in some Latin American states, for instance, the constitutions of Bolivia and Ecuador (Gudynas, 2011; Acosta and Abarca, 2018).

In common with other radical accounts of justice, postcolonial theories and approaches offer important insights into the ways in which humans are *social* beings, how power – and especially the power inherent in centre–periphery relationships – *unevenly* shape human life and its possibilities, and how these centre–periphery relations have been as essential to the colonisation of nature as colonisation of humans. Postcolonial thinking offers, in this respect, a powerful *critique* of injustice, while also pointing, on the one hand, towards the need for rectificatory processes and practices, while not, on the other hand, becoming a prisoner of the past (as Fanon warned against). Precisely *how* this should be accomplished, however, is something that scholars positioned in postcolonial theory have thus far remained largely silent about. Nonetheless, the value of postcolonial *critique* for assessing the justness – or not – of various processes ought to be clear. This critique asks us to look out not only for maldistribution, but also for whether epistemic injustice might be being perpetuated under the guise of, for instance, sustainability.

Notes

[1] On the scope, range, and *location* of postcolonial thinking see Runesson (2011), Loomba (2005), Ashcroft et al (2002) and Bhabha (2004 [1994]). On the degree to which postcolonialism has developed disciplinary ambitions, see Lazarus (2004). For an introduction to the entwined history of the relationship between postcolonialism, postmodernism and the rise of cultural studies, see Quayson (1998).

[2] While also raising the question of whether such subalterns can, within the confines of ongoing European hegemony, even speak in the first place, much less be heard (Spivak, 1999 [1988]).

[3] As in feminism and critical social theory, the question of recognition is paramount. See Chapters 4 and 5.

[4] In this it has much in common with feminist theories of justice, see Chapter 4.

[5] The major exception is Bartels et al (2017), which we will discuss in one of the following sections.

[6] This also goes for most of libertarianism, but is to a larger degree questioned in cosmopolitanism.

[7] Theories of epistemic injustice cannot be fully equated with postcolonial thinking, but the two fields do have intersecting and, in some ways, mutually supportive histories. There is a good deal of crosspollination between them. More typically, theories of epistemic injustice are categorised as a kind of *post-structural* theory, but the ties that bind postcolonial with post-structural theorising are tight, and it is neither the goal of this chapter, nor necessarily helpful to it, to seek to untie those binds here. In addition, Fricker is positioned firmly within feminist thinking.

[8] Spivak's scholarship is also positioned within feminist thinking, again indicating the connectedness between feminist and postcolonial thinking and critique.

[9] In the *Prison Notebooks*, Gramsci (1971) used 'subaltern' as a code for any class of people (but especially peasants and workers) subject to the hegemony of another, more powerful class of people.

[10] First published in 1983.

[11] A decade after writing about the nature of this encounter in *Black Skin, White Masks* (1967 [1952]), Fanon recognised in *The Wretched of the Earth* (1963 [1961]) that transforming its structural conditions might necessarily be a violent process.

[12] It is in this sense that calls to 'decolonise the academy' should be understood: they are calls to transform the distribution of recognition, both in relation to the question of who produces valid knowledge and where it is produced.

[13] By 'mimicry' Bhabha meant the ways in which colonised peoples imitate their coloniser's culture.

[14] There are affinities here with James Scott's (1985) theorisation of the 'weapons of the weak' as vital components in the relations and institutions of power that structure coloniality and postcoloniality.

[15] Interestingly, given Chakrabarty and other's ruling out of Kantian or other influences for postcolonial theorising, Spivak's argument here is pure Kantian idealism (as Wainwright concedes), recapitulating Kant's concept of the *hiatus irrationalis* which intervenes between the phenomenon and the thing-in-itself.

[16] See also, in this regard, Iris Marion Young's discussion of the relation between *liability* and *responsibility* outlined in Chapter 4, on feminist approaches to justice.

[17] These are vital questions in the contemporary era and drive the work of most 'truth and reconciliation' processes set up in the wake of significant political transformation (as in South Africa after apartheid), genocide (as in Rwanda and Burundi), and moments of significant political violence (as in the truth and reconciliation process in Greensville,

South Carolina, USA, seeking rectification in the wake of a historical race-riot by Whites against African Americans: see Inwood [2012]).

[18] Spivak (in Hedge and Shome, 2002, p 272), in her typically oracular way, seeks to negotiate a middle ground between dwelling in an imprisoning past and fighting for a liberating future: 'In order to give globalization historical depth you must move it to postcoloniality', though what this really means is left open.

References

Acosta, A. and Abarca, M.M. (2018) Buen vivir: An alternative perspective from the peoples of the global south to the crisis of capitalist modernity. In V. Satgar (ed), *The Climate Crisis: South African and Global Democratic Eco-Socialist Alternatives*. Johannesburg: Wits University Press, pp 131–147.

Aristotle (1999) *Nicomachean Ethics*, translated by T. Irwin. Indianapolis, IN: Hackett.

Ashcroft, B., Griffiths, G. and Tiffin, H. (2002) *The Empire Writes Back: Theory and Practice in Post-colonial Literatures*, 2nd edn. London: Routledge.

Ashcroft, B., Griffiths, G. and Tiffin, H. (2007) *Post-colonial Studies: The Key Concepts*. London: Routledge.

Bartels, A., Eckstein, L., Waller, N. and Wiemann, D. (eds) (2017) *Postcolonial Justice*. Leiden and Boston: Brill, Rodopi.

Bhabha, H.K. (2004 [1994]) *The Location of Culture*. London: Routledge.

Chakrabarty, D. (2012) Postcolonial studies and the challenge of climate change. *New Literary History*, 43(1), 1–18.

Collste, G. (2015) *Global Rectificatory Justice*. Hampshire, UK: Palgrave Macmillan.

Cornell, D. (2010) The ethical affirmation of human rights: Gayatri Spivak's intervention. In R. Morris (ed), *Can the Subaltern Speak? Reflections on the History of an Idea*. New York: Columbia University Press, pp 100–114.

Dirlik, A. (1994) The postcolonial aura: Third world criticism in the age of global capitalism. *Critical Inquiry*, 20(2), 328–356.

Fanon, F. (1963 [1961]) *The Wretched of the Earth*. New York: Grove Press.

Fanon, F. (1967 [1952]) *Black Skin, White Masks*. New York: Grove Press.

Foucault, M. (1980) *Power/Knowledge: Selected Interviews and Other Writings, 1972–77*. New York: Vintage.

Fricker, M. (2007) *Epistemic Injustice: Power and the Ethics of Knowing*. New York: Oxford University Press.

Fricker, M. (2013) Epistemic justice as a condition of political freedom. *Synthese*, 190, 1317–1332.

Fricker, M. (2018) Epistemic injustice and recognition: A new conversation – afterword. *Feminist Philosophy Quarterly*, 4(4), 1–5.

Gathii, J.T. (2011) TWAIL: A brief history of its origins, its decentralized network, and a tentative bibliography. *Trade, Law and Development*, 3(1), 26–64.

Gramsci, A. (1971) *Selections from the Prison Notebooks of Antonio Gramsci*. New York: International Publishers.

Gudynas, E. (2011) Buen vivir: Today's tomorrow. *Development*, 54(4), 441–447.

Hegde, R.S. and Shome, R. (2002) Postcolonial scholarship – productions and directions: An interview with Gayatri Chakravorty Spivak. *Communication Theory*, 12(3), 271–286.

Inwood, J.F. (2012) Righting unrightable wrongs: Legacies of racial violence and the Greensboro Truth and Reconciliation Commission. *Annals of the Association of American Geographers*, 102(6), 1450–1467.

Ismael, Q. (2005) *Abiding by Sri Lanka: Peace, Place and Postcoloniality*. Minneapolis, MN: University of Minnesota Press.

Kohn, M. and Reddy, K. (2017) Colonialism. In E.N. Zalta (ed), *The Stanford Encyclopedia of Philosophy*. https://plato.stanford.edu/archives/fall2017/entries/colonialism/

Lazarus, N. (2004) Introducing postcolonial studies. In N. Lazarus (ed), *The Cambridge Companion to Postcolonial Literary Studies*. Cambridge, UK: Cambridge University Press, pp 1–16.

Loomba, A. (2005) *Colonialism/Postcolonialism*, 2nd edn. New York: Routledge.

McGonegal, J. (2009) *Imagining Justice: The Politics of Postcolonial Forgiveness and Reconciliation*. Montreal: McGill-Queen's University Press.

Mignolo, W.D. and Walsh, C.E. (2018) *On Decoloniality: Concepts, Analytics, Praxis*. Durham, NC: Duke University Press.

Mohanty, C.T. (2003) *Feminism without Borders: Decolonizing Theory, Practicing Solidarity*. Durham, NC: Duke University Press.

Mutua, M. and Anghie, A. (2000) What is TWAIL? *Proceedings of the Annual Meeting (American Society of International Law)*, 94, 31–40.

Nair, S. (2017) Introducing postcolonialism in international relations theory. In S. McGlinchey, R. Walters and C. Scheinpflug (2017) *International Relations Theory*. Bristol: E-International Relations.

Plumwood, V. (2003) Decolonizing relationships with nature. In W.M. Adams and M. Mulligan (eds), *Decolonizing Nature: Strategies for Conservation in a Postcolonial Era*. London: Earthscan, pp 51–78.

Quayson, A. (1998) Postcolonialism. In *Routledge Encyclopedia of Philosophy*. London: Routledge.

Roberts, R.C. (2011) Rectificatory justice. In D.K. Chatterjee (ed), *Encyclopedia of Global Justice*. Dordrecht, The Netherlands: Springer, pp 936–938.

Roberts, R.C. (2017) Race, rectification, and apology. In N. Zack (ed), *The Oxford Handbook of Philosophy and Race*. New York: Oxford University Press, pp 516–525.

Robinson, J. (2005) *Ordinary Cities: Between Modernity and Development*. London: Routledge.

Runesson, A. (2011) The theoretical location of postcolonial studies. In *Exegesis in the Making*. Leiden, The Netherlands: Brill, pp 17–50.

Said, E.W. (1978) *Orientalism.* New York: Pantheon Books.

Said, E.W. (1993) *Culture and Imperialism.* New York: Knopf/Random House.

Scott, J.C. (1985) *Weapons of the Weak: Everyday Forms of Peasant Resistance.* New Haven, CT: Yale University Press.

Spivak, G.C. (1999 [1988]) Can the subaltern speak? In *A Critique of Postcolonial Reason: Toward a History of the Vanishing Present.* Cambridge, MA: Harvard University Press, pp 66–111.

Spivak, G.C. (2012) *An Aesthetic Education in an Age of Globalization.* Cambridge, MA: Harvard University Press.

Turnbull, D. (2017) Postcolonial injustice: Rationality, knowledge, and law in the face of multiple epistemologies. In A. Bartels, L. Eckstein, N. Waller and D. Wiemann (eds), *Postcolonial Justice.* Leiden and Boston: Brill, Rodopi, pp 3–16.

Wainwright, J. (2013) *Geopiracy: Oaxaca, Militant Empiricism, and Geographical Thought.* New York: Palgrave Macmillan

Young, I.M. (2000) *Inclusion and Democracy.* Oxford, UK: Oxford University Press.

Young, I.M. (2011) *Responsibility for Justice.* Oxford, UK: Oxford University Press

Young, R.J.C. (2012) Postcolonial remains. *New Literary History*, 43, 19–42.

7

Indigenous Approaches to Justice

Stephen Przybylinski and Johanna Ohlsson

Introduction

In this chapter, we provide an overview of Indigenous perspectives on justice. As non-Indigenous scholars ourselves, our survey does not develop arguments by adopting Indigenous perspectives, nor does it attempt to speak for Indigenous peoples. Rather, the chapter draws from a breadth of legal, political and normative Indigenous scholarship to illustrate the contributions of Indigenous peoples and scholars to understandings of justice. We hope that this may contribute not only to challenging the continuing marginalisation of such perspectives within justice theorising (Watene, 2020), but also to show why such perspectives advance justice theorising more broadly, illustrating the ways in which these perspectives help us rethink dominant or mainstream positions on justice.

Defining exactly what 'Indigeneity' means, and who 'Indigenous peoples' are, is challenging. Indigenous peoples cannot be reduced into a singular group of simply Indigenous *people*, as Indigenous peoples vary in their geographies, cultures, languages, and social and political institutions (Sarivaara et al, 2014). But it also risks reinforcing exclusionary boundaries of those definitions by more dominant groups (Capeheart et al, 2007). The term Indigenous can be more or less inclusive depending on national legislation and how specific states or organisations categorise their demographic data which is also contingent on whether people self-identify as Indigenous. Problematically, this identification has often been connected to various stigma, often due to the discrimination many Indigenous groups have experienced.

Nevertheless, many suggest some definition of Indigenous peoples or groups is necessary. Without a legal term to which Indigenous peoples can appeal, for example, states will not, for better or worse, grant Indigenous groups far-reaching rights (Scheinin, 2005, p 13). While legal rights are not the only means by which justice may be worked towards, as this chapter will show, rights are one of the most straightforward means by which Indigenous justice claims are respected. With this in mind, different definitions of Indigenous or Indigeneity exist from organisations such as the United Nations (UN) and the International Labour Organization (ILO). The UN has not adopted an official definition (indicating the politically sensitive nature of doing so), but generally speaking, it states that Indigenous peoples are 'descendants ... of those who inhabited a country or geographical region at the time when people of different cultures or ethnic origins arrived' (UNPFII, 2007, p 1). The ILO by contrast defines Indigenous peoples as descendants 'from the populations which inhabited the country, or a geographical region to which the country belongs, at the time of conquest or colonisation or the establishment of present state boundaries and who, irrespective of their legal status, retain some or all of their own social, economic, cultural and political institutions' (International Labour Organization, 2017, np). Such definitions speak to historical and contemporary relations between Indigenous peoples with settler–coloniser states and in themselves point to key issues regarding claims for justice.

The chapter begins by identifying how Indigenous ontologies and epistemologies generally shape understandings of justice. Indigenous scholars underscore relationships with the natural and sometimes sacred world. As such, a particular emphasis on maintaining harmony with human and non-human life is represented in many Indigenous understandings of justice. After identifying ontologies and epistemologies of justice, the chapter moves on to examine a larger body of political theory, which details common normative and legal claims of Indigenous peoples within the context of western notions of justice. A main instrument through which Indigenous peoples make claims for justice has been through an appeal to rights, both legally, politically and morally. As such, we highlight three types of rights which are predominant throughout this literature: rights to ancestral lands and resources; rights to self-determination; and rights to cultural preservation. While a rights-based notion of justice for Indigenous peoples is largely presented as the recognition (that is, respect for and implementation) of rights under common law, debates question whether legal rights can accommodate Indigenous ways of knowing within common law. In ending, we address why rights specific to Indigenous peoples matter for advancing claims of justice, but also why such claims emanate from the relationality between ontological and epistemological understandings of what is just.

Indigenous ontologies and epistemologies

Before conceptions of Indigenous justice are articulated, it is useful to have more context on how Indigenous scholars describe the relationality between ways of knowing (epistemology) and being (ontology). In general, Indigenous ways of knowing and being centre on the relationship between the physical, human and sacred worlds. As one Indigenous scholar puts it:

> The physical world is the base that is land, the creation. The land is the mother, and we are of the land. We do not own the land, the land owns us. The land is our food, our culture, our spirit, our identity. The physical world encapsulates the land, the sky, and all living organisms. The human world involves the knowledge, approaches to people, family, rules of behavior, ceremonies, and their capacity to change. The sacred world is not based entirely on the metaphysical, as some would believe. Its foundation is in healing (both the spiritual and physical well-being of all creatures), the lore (the retention and reinforcement of oral history), care of country, the laws and their maintenance. (Foley, 2003, p 46)

Sustaining harmonious relationships between all three elements – physical, human, sacred – is important as it would be difficult to isolate one element as being more important than another. Such a perspective lends itself to thinking about justice relationally and holistically.

Given the anthropocentrism of most western philosophies and theories of justice, it is worth emphasising how relations with the natural world are central to Indigenous ways of knowing and being. For many Indigenous peoples, there is no distinction between human and nature as one's environment is seen as kin (Caillon et al, 2017). Natural relations are often understood as relations with all forms of life; humans, animals, and all other biophysical features of Earth can be understood and valued as relatives in relation to one another (McGregor, 2018). Kin relationships in turn inform some Indigenous ontological positions on environmental decision-making. As McGregor (2018 p 16) notes, in many Indigenous legal traditions, 'humans alone may not be the focus or even the architects of [Indigenous] laws; the universe can be seen as having innate laws for governing itself in moral and appropriate ways. In this view, humans alone do not create law, nor in some cases are they responsible for enforcing law'. Such a position points to a potential connection to ideas within the natural law tradition, but Indigenous positions more clearly stress environmental aspects, moving beyond the anthropocentric focus commonly seen in natural law. Indigenous ontological and epistemological positions underscore the relationality of the natural world, informing Indigenous perspectives in general and on justice in particular, which aim to balance the physical, human and sacred worlds.

Given a more metaphysical approach than that of much of western philosophy and normative approaches to justice, the exclusion of Indigenous epistemologies (and ontologies) from mainstream justice theorising constitutes one form of injustice in itself. Addressing the perceived gap between Indigenous and western epistemologies, Lesley Le Grange and Carl Mika (2018, p 503) argue that western philosophy traditionally 'has been premised on the insertion of a unit of representation between the self and the thing being discussed, so that it can be discussed objectively'. A more metaphysical approach to Indigenous philosophy, they argue, 'does not distinguish between the idea of something and the thing itself' (Le Grange and Mika, 2018, p 502). Instead, 'the entities that comprise the creation stories of various [Indigenous] groups are at once both [capable of being entities] and able to be represented. They are present whilst being conceptualized and discussed' (Le Grange and Mika, 2018, p 502). In this way, Indigenous ways of knowing can be in tension with western epistemological approaches, the result of which could be seen as *epistemic* injustice. Epistemic injustice presents as invalid any 'forms of knowledge that differ from dominant rationality ... [which is] manifest, among other things, in the common assumption that modern science is objective and universal, while indigenous forms of knowing are not credible' (Widenhorn, 2014, p 378). Advancing epistemic justice, therefore, means to value Indigenous knowledge as legitimate and equivalent to scientific assumptions.

Given the marginalisation of Indigenous ways of knowing within western philosophy in general, it is unsurprising that Indigenous approaches to justice specifically have remained largely outside of dominant approaches to justice theorising. Drawing from Indigenous scholars, in the following section we detail how these ways of knowing have been understood within Indigenous approaches to justice and how such approaches to justice are articulated in relation to mainstream theories of justice.

Indigenous approaches to justice

Central to Indigenous conceptions of justice is the process of healing. There are at least two ways to understand justice as healing. One is practised as a process of conflict resolution and reconciliation among members within Indigenous communities, a process conducted separate from western legal and criminal justice systems. The Maori Indigenous peoples of New Zealand, as Capeheart and Milovanovic (2007, p 109) show, respond to community harms through healing, often embracing an elder's evaluations of 'concrete situations' of conflict instead of by applying a systematised process of formal law. Here, justice is advanced as a way of healing and promoting relationships among community members by requiring a collective responsibility in finding this balance. As Vieille (2012, p 9) notes, 'implicit in the sense of collective responsibility that underpins relational justice is the belief that all community members are responsible for sustaining community well-being. Justice, therefore, serves as a means of maintaining that harmony or balance within the community'. This makes justice an inherently social concept. However, justice is not only about maintaining harmony *between* individuals involved in conflict. Maori justice more broadly seeks to heal 'the person who has wronged, the person who has been wronged, and the entire network of relationships affected by the harm caused' (Vieille, 2012, p 10). In this way, Vieille argues, justice is carried out in the public sphere, 'as a public space for all [within the community] to participate'. Further, restorative justice may extend to healing wrongs not only between humans but also between relatives in its broader metaphysical understanding.

A second form of justice as healing happens as a response to Indigenous communities' emotional and material traumas of colonialism and settler colonisation. Acknowledging and accounting for legacies of oppression against Indigenous peoples requires more than rights to resources alone and ought to recognise how such oppressions are reinforced today. The violence of land dispossession, broken treaties, forced displacement and schooling of Indigenous children, and loss of cultural expressions and language, for instance, are not events resigned to the past but are ongoing oppressions requiring healing. Generations of trauma for Indigenous peoples brought on by settler states requires that a more substantive process of healing as justice supplement justice beyond the advancement of rights. As Watene (2016a, p 139) notes, healing for Indigenous peoples 'starts with recognizing these events and conditions, and acknowledging the way that they impact on their lives'. To publicly acknowledge traumatic events, she notes, can 'open up space for indigenous communities to remember, face, and begin to overcome histories of grave injustice and great loss of hope' (Watene, 2016a, p 139). Healing affords one procedural mechanism through which Indigenous peoples can move forward from the injustices of colonisation.

Restorative justice may be practised with a goal of finding harmony within communities, but Indigenous peoples are forced to navigate the constraints of legal systems of dominant societies at the same time. Given that much discourse on Indigenous justice as contextualised through western notions of justice centres on rights (see section 'Perspectives on Indigenous rights'), little scholarship has articulated where Indigenous epistemological approaches to justice specifically and explicitly converge with western conceptions of justice. One of the few integrations of Indigenous approaches to justice theorising with those of western theories of justice is found in the work of Krushil Watene. A Maori scholar, Watene (2020, p 166) denounces the absence of Indigenous perspectives in mainstream justice theorising, arguing that Indigenous peoples remain not only 'unable to define who they are and want to be', but who are also 'denied an active role in [their] (social, political, economic, cultural, and environmental) conceptual landscapes'. Watene argues that such epistemic injustice requires that Indigenous philosophers be incorporated into mainstream justice theorising so that Indigenous approaches to justice can be appropriately articulated.

Watene's contribution to justice theorising has primarily been through critical reflection on the Capabilities Approach (CA; see Chapter 8). Watene argues that, compared with many other theories of western liberal justice, the CA has the advantage of being flexible as a methodology (as well as for its rejection of ideal principles). In particular, Watene values its potential focus on people's capabilities or opportunities to live a life they have reason to value, an openness she believes underscores a sense of self-determination central to Indigenous claims to sovereignty. Hence, Watene's contribution both challenges and develops the CA.

Watene is critical of the CA for at least two reasons. The first, she notes, concerns how the CA focuses on the individual. Here, well-being is understood as the capability of an individual to flourish as they themselves best see fit. With such a singular focus, the Indigenous significance placed on community relationships is lost. In that kin relationships are central to Indigenous ways of being and knowing, Watene (2020, p 169) notes that the idea of community is also 'intimately connected to place', so much so that lands and natural resources 'provide the context for indigenous peoples' values [which] are crucial for revitalizing, reproducing, and maintaining indigenous ways of being, knowing, and doing'. The inability of CA to recognise collective and cultural aspirations as indicators of well-being in turn undermines Indigenous ontologies.

Following closely from the first critique, Watene argues that, not only is the CA too individually focused, but that the focus remains on humans alone. Critical of how Amartya Sen and Martha Nussbaum address the concept of 'nature', Watene (2016b, p 291) argues that, for them, 'the natural world is not valuable in its own right, but only valuable in light of the importance of either human agency or dignity'. To not value nature or the natural world as kin itself is problematic. The danger in not valuing nature for its own sake, as Watene argues Sen and Nussbaum do, means that 'natural resources can be substituted – particular lands, waterways and resources need not be preserved as such, but [hold] only the general capacity to create well-being' (2016b, p 293). From the CA perspective, the extent to which resources are depleted or destroyed matter only in how they affect human lives. Thus, to exclude Indigenous views on land and resources, as kin relations, renders insignificant a collective means of understanding well-being when seen through the CA.

To better incorporate Indigenous approaches into CA requires recognising collective forms of well-being, and most importantly through values identified by Indigenous communities themselves. Yet, as Watene (2020, p 173) acknowledges, it is one thing to include 'indigenous philosophies in [mainstream] justice theorizing', but it is entirely another to suspend epistemic barriers so that Indigenous peoples may theorise justice 'on their own terms'. But herein lies a dilemma regarding the marginal position of Indigenous perspectives on justice. How can Indigenous perspectives become incorporated into western-dominated approaches to justice *without* reducing and misrecognising the significance of Indigenous ways of knowing? For instance, how can a *collective* understanding of justice, one which sees the natural world as a set of relationships beyond those of humans alone, figure into a liberal paradigm premised solely on human individuality? As Capeheart and Milovanovic (2007, p 117) ask, *must* 'indigenous peoples, in seeking redress to grievances, express themselves in the written form of dominant groups and thus subject themselves to a transformation of their way of life?' The obvious answer is no; Indigenous peoples should not reduce their ways of knowing by adjusting to liberal frameworks of justice alone. For, to do so risks excluding non-dominant conceptions of justice whose values derive outside of western liberal 'rationality'. It is this dilemma which seems to frame a larger discourse on Indigenous justice, a discourse centring on how and whether liberalism can enable Indigenous recognition and self-determination as expressed through rights discourse. We turn to detail the rights-based approach to justice in the following section.

Rights-based approaches to Indigenous justice

Much literature on Indigenous perspectives on justice centres on rights. To recognise and respect rights for Indigenous peoples has been a primary way to address historic and contemporary injustices between nation states and Indigenous groups. For, the injustices that Indigenous peoples experience today must be understood as transgenerational processes of colonisation, political disenfranchisement and a denial of cultural diversity by dominant society (Kuppe, 2009, p 103). The social movement and principle of *buen vivir* or 'living well', for example, attempts to decolonise the economic and social relations underscoring unchecked extractivism of capitalist development by establishing rights frameworks grounded in Indigenous conceptualisations of human and non-human relationships with the environment (Acosta and Martinez Abarca, 2018). But more generally, the Indigenous rights literature focuses on the specific rights types of land rights, rights of self-determination and to culture, which accord with the aforementioned injustices. The following section details how these specific rights are comprised before addressing whether Indigenous rights differ from other categories of rights. Doing so illustrates why rights are significant for the advancement of specifically Indigenous justice.

Like most rights struggles, Indigenous peoples' rights claims emanate from situated experiences of oppression. Nonetheless, Indigenous rights claims often share more general characteristics. For example, the United Nations

Declaration of the Rights of Indigenous People (UNDRIP), adopted in 2007, is regularly cited because of the way in which it represents Indigenous struggles for recognition. Though not legally binding, the document is nevertheless an important contribution towards increased recognition of Indigenous peoples' rights and freedoms. Throughout the 46 articles in UNDRIP (United Nations, 2008), there exist a set of rights categories which may be broadly grouped as: rights to land; rights to self-determination; and rights to cultural preservation.[1] The *rights to land* expressed in Articles 10, 26, 28, 29 and 32 call for the right of Indigenous peoples to not be forcibly removed from their lands or territories; their right to own, use, develop, control lands and territories; and a right of redress or restitution for lands and territories confiscated without consent. The *rights to self-determination* expressed in Articles 3, 4, 5, 33, 35 and 37 call for the right of Indigenous peoples to determine their political status; the right to autonomy or self-government in matters regarding internal affairs; the right to maintain and strengthen political, legal, economic, social and cultural institutions, while retaining the right to participate fully in the life of the state; and the right to recognition, observance and enforcement of treaties and other agreements with states. Finally, *rights to cultural preservation* expressed in Articles 8, 11, 12 and 15 call for Indigenous peoples' right to not be subject to forced assimilation or cultural destruction; the right to practice and revitalise cultural and spiritual traditions, customs and ceremonies; and the right to culturally specific education and language preservation. These categories of rights are not mutually exclusive or exhaustive. Rather, they reflect elements common to Indigenous rights literatures more broadly and offer a source of political leverage.

Perhaps the most prominent rights among these three is the right to land. Given the historical seizure by settler states of Indigenous peoples' ancestral lands, recognising rights to landed resources and control over territory are tangible actions that states can take to correct for historic injustices. But land itself is not only about control over resources. Land factors prominently into Indigenous ontologies and epistemologies about the nature of life. As J.M. Valadez (2012, p 699) notes, 'most indigenous groups view themselves as connected to their ancestral past through continued stewardship of the land, and they consider caring for the land that future generations will inhabit as a great moral and cultural responsibility'. To have land returned is not simply being able to benefit from resources, returning stolen property acknowledges the cultural value of Indigenous connection with traditional lands.

The right to traditional lands in itself does not encompass all that Indigenous peoples have lost through colonisation. Indeed, the act of returning land in itself cannot achieve meaningful redress, as Douglas Sanderson argues:

> [B]ecause lands alone do not achieve the correlative requirements of corrective justice. To make redress correlative it is necessary to return

to Indigenous people the full scope of what it is that was taken from them and that is today still being denied to them; namely, the right to create and maintain a set of institutions that positively affirm and promote Indigenous identity in Indigenous communities. (Sanderson, 2012, p 131)

Closely connected with a right to land, therefore, is the right of self-determination.

Self-determination embodies the concepts of self-government and sovereignty, although Indigenous claims for self-determination rarely extend as far as independent sovereign statehood itself (Ivison et al, 2000, p 14). Instead, claims for recognising Indigenous sovereignty tends to concentrate on 'the design of institutions, the making of laws and local constitutions, the management and land of resources, and the determination of strategies for community and economic development' (Patton, 2019, p 1266). Further, self-determination allows 'not only political governance that accords with the will of indigenous peoples but also governance that is consistent with their historical modes of sociopolitical organization' (Valadez, 2012, p 699). For, what Indigenous peoples lost through historical acts of injustice, Sanderson (2012, p 129) notes, is 'the right and capacity ... to order their lives and to exercise their freedoms in particularly Indigenous ways'.

Unfortunately for many Indigenous groups, the freedom and power to shape their own futures is dependent on nation states. Since sovereignty is largely dependent on nation states to respect Indigenous rights claims, claims for self-determination become contested. When nation states contest Indigenous claims for self-determination, Indigenous groups are forced into making rights claims within the political and legal systems of a given nation state. Cornell frames this as 'positional politics'. Positional politics concerns the standing of collectives within 'an encompassing political system and secondarily, within the encompassing economy' (Cornell, 2018, p 14). Given this, Cornell notes two dimensions of Indigenous approaches to justice: justice as position and justice as practice. He states that justice as position

refers to the position that Indigenous peoples ... occupy within encompassing political systems. The key issues in justice as position are *recognition* (are Indigenous peoples recognized as formal political actors – governments – with whom central government should interact on a government-to-government basis?), *jurisdiction* (what is the nature and scope of Indigenous jurisdiction over space, persons, relationships, and activities?), *power* (regardless of the *de jure* nature of jurisdiction, what is its *de facto* nature?), and *organizational freedom* (to what extent are those nations free to organize as they see fit in pursuit of collectively determined goals?). (Cornell, 2018, p 15, emphasis added)

To the extent that these four elements increase, so too does justice as position increase. What is unjust for Cornell is not only that colonialism 'leads to the expropriation of land, labour, natural resources, intellectual property, and so forth- but also ... its denial of justice as position' (2018, p 15). Significantly, however, an increase in justice as position leads to a new problem: justice as practice. 'The empowerment of indigenous peoples changes their relationship to justice', he notes. With 'recognition, jurisdiction, power, and organizational choice, outcomes such as fairness, accountability, justice ... becomes an indigenous nation's own responsibility' (Cornell, 2018, p 15). For Cornell, justice as practice is the exercise in this responsibility. Cornell argues that nation states facilitate justice as position through certain rights for Indigenous groups, but do not assist groups in achieving justice as practice. That is, nation states do not enable Indigenous groups' real self-governing power on Indigenous terms, but, rather, through institutional practices common to liberal democracies.

Finally, rights to cultural preservation are indispensable for Indigenous claims to justice. The right to cultural integrity is important for different reasons. The maintenance of languages, religious practices and customs not only hold value in themselves, but rights which preserve cultural practices reinforce Indigenous peoples' self-determination. Importantly, the preservation of Indigenous cultural traditions cannot be carried out by isolated individuals. It requires there to be communities through which traditions can be practised (Valadez, 2012). In other words, Indigenous groups must be recognised and respected as self-determining peoples for such cultural rights to be respected and justice to be pursued.

Some of these categories of rights are highly relevant also for non-Indigenous peoples, such as other minority groups. But what defines a specifically *Indigenous* right? The next section details debates over how such rights have been understood.

Perspectives on Indigenous rights

What are *Indigenous* rights? Are they necessarily legally codified rights in liberal states? If not, what makes them distinct from liberal rights generally? These questions have been pervasive in debates over Indigenous rights and claims for justice. Three debates in particular provide context on Indigenous rights and their relationship to liberalism. The first concerns whether they are a species of human rights or if they go beyond this anthropocentrism including some other rights type, the second concerns the difference between individual and group rights, while the third regards the difference between Indigenous and minority status.

Indigenous specific rights are meant to protect the interests of 'culturally distinctive groups with ancestral connections to precolonial peoples who inhabited areas now occupied by settler societies' (Valadez, 2012, p 696). Such rights recognise and address the needs of Indigenous peoples particularly in relation to the dominance of non-Indigenous societies. Ivison (2003) notes two ways of conceptualising Indigenous rights. The first observes the 'historical, cultural and political specificity of the interests to which the claims appeal – in other words to indigenous difference' (Ivison, 2003, p 325). The other approach is an 'appeal to general or human rights' which argues that Indigenous rights are a 'species' of human rights, in that they refer to the general rights and legitimate claims everyone has a right to have. From either perspective, Indigenous rights are most often understood as 'common law rights – as legal rights … that emerge from a complex of cross-cultural practice of treaty-making – and as being already inside and thus enforceable by the law'. This indicates a legalistic understanding of rights, as rights are primarily put to force when codified in law broadly understood.

Given the common law recognition of Indigenous rights, debate exists over whether Indigenous rights are and ought to be distinct from other rights. Arguments against rights specifically for Indigenous groups note the challenge in recognising a group's collective rights. In this view, group or collective rights are said to detract from liberal values of equality and universalism.[2] In that Indigenous group rights have culturally specific interests recognised, so this argument goes, such rights violate the essentially individualistic nature of liberal moral individualism (Ananya, 1999). In other words, appeals to strictly liberal egalitarianism demand a uniform system of rights and responsibilities undifferentiated by race, ethnicity, religion, class, and so on (Patton, 2019). Equality in this sense means that no one group receives special recognition, which risks failing to account for historical injustices as well as current power structures.

The argument against group rights is not ubiquitous in rights theories, however. Others see group rights as a primary way in which Indigenous peoples may overcome inequities within settler-colonial states. As Ivison (2003, p 335) notes, Indigenous rights acknowledge 'the formal equality of peoples who were previously considered in both international and domestic law as politically (and culturally) inferior and thus undeserving of equal consideration'. Functionally, then, group rights enable 'indigenous groups to address the social and economic disadvantages they suffer from, taking into account their unique historical circumstances' (Ivison, 2003, p 335). Particularly since rights are continuously in need of justification, rights specific to Indigenous peoples function as a reminder of why the interests of Indigenous groups warrant protection. Indigenous rights thus express not only legal instruments but also normative positions reflecting how such groups are able to overcome past and contemporary injustices in relation to dominant societies.

Individualist arguments against collective rights often point to the incommensurability of interests represented between individuals and groups. Yet, all rights can be incommensurable with one another, including individual rights themselves. As such, if it is possible to morally justify individual rights, so too can moral foundations for group rights be justified. It is possible to justify collective rights, Dwight Newman (2007) argues, because 'it is possible to identify certain group interests – things that make a group's or community's life go better, that make the community thrive and flourish – that are irreducible to individual interests whose fulfillment is at the same time a precondition to the meaningful realization of individual interests that ground rights' (Newman, 2007, p 281). Given that all rights can be incommensurable, moral reasons for recognising legal rights of Indigenous peoples differ none in this regard.

That is not to say that claims for cultural recognition are easy to determine in the form of legal rights. Nation states often disagree on what constitutes group interests, but also often fail to be fully inclusive when identifying the group. Discourse on 'cultural rights' help to frame how we can think about Indigenous rights specifically within the uneven power relations between Indigenous groups and states, minority and majority groups within a state, or even within minority or Indigenous groups. From a liberal culturalist perspective, such as that of Will Kymlicka, Indigenous groups are vulnerable to the decisions of the dominant society around them. To remedy the injustices of Indigenous groups by settler societies, Kymlicka argues, 'requires not identical treatment but rather differential treatment in order to accommodate differential needs' (1995, p 113). This clearly brings to the fore the notion of justice. For, 'group-differentiated self-government rights compensate for unequal circumstances which put the members of minority cultures at a systemic disadvantage in the cultural market-place' (Kymlicka, 1995, p 113). At the same time, the liberal culturalist argument for differential group rights may potentially misrepresent the historic injustices perpetrated by colonial states. As Patton (2019, p 1269) notes, claims for group rights rest 'on the need for equal access to cultural membership on the part of individual members of minority cultural groups'. What may be overlooked in culturally distinct rights appeals, therefore, is Indigenous claims to a fuller extent of rights, such as to land (and resources) and self-determination, rights which are central to many Indigenous groups' claims for recognition.

Kymlicka's earlier arguments on differential rights are not explicitly centred on Indigenous peoples, but reflect on rights of minority groups more generally. He has since turned attention to the distinction between minority groups and Indigenous peoples. And in doing so, he explicates further reasons for Indigenous-specific rights by noting the difficulty in justifying how Indigenous claims compare with those of other minority groups. He notes the distinction the UN makes between 'integration-seeking' minority groups and 'autonomy-seeking' Indigenous groups obscures the ability to easily differentiate between 'historically-settled homeland minorities' who have similar claims to territory and self-government as Indigenous peoples (Kymlicka, 2008, p 7).[3] Further nuancing these distinctions is beyond the point of this chapter. What is significant here is that Indigenous peoples often have relatively more defined sets of claims for rights which other minority groups do not always share. As Weigård (2008, p 177) argues, 'indigenous peoples around the world have long made claims not just to be treated as other ethnic minorities, but also that they should be given a special status, firstly because they are *peoples/nations* and secondly because of their indigenousness or aboriginality' (emphasis in original). Such claims simultaneously seek recognition of historical land rights along with political self-determination.

Rights thus hold value for Indigenous claims to justice for different reasons. Indigenous claims to rights recognise and hold accountable the unjust historical actions of settler states and facilitate legal relations within the state apparatus so that culturally specific claims can be advanced and respected. Rights also hold normative value for Indigenous groups searching for more just relations with dominant society. In this way, rights of self-determination and to ancestral lands that are often recognised by liberal states may help promote Indigenous approaches to justice, ways of expressing culturally specific understandings about what just relationships look like that are not as easily captured in a legal, common law framework.

How Indigenous approaches to justice relate to applied fields of justice

Given the prominence of natural and physical relations to Indigenous ways of knowing and being, it is unsurprising that Indigenous perspectives have been represented in the environmental justice (EJ) literature. For instance, Indigenous knowledge is often conflated as the same thing as 'traditional ecological knowledge' in the scholarship on environmental justice (Pierotti, 2010; Kim et al, 2017). But Indigenous justice need not necessarily be equated as Indigenous *environmental* justice. While the EJ literature recognises environmental knowledge of Indigenous peoples, perspectives about specifically Indigenous justice are less prominent within EJ literatures. That is, EJ can often lack insights of Indigenous epistemologies of justice. Such are the arguments of Anishinaabe scholar Deborah McGregor (2018, p 9), who advocates that Indigenous peoples 'move beyond "Indigenizing" existing EJ frameworks and seek to develop distinct frameworks that are informed by Indigenous intellectual traditions, knowledge systems, and laws'. McGregor notes, for example, that the Anishinaabe concept of *Mino-Mnaamodzawin* or 'living well' embraces a relational approach to living on the Earth whereby human and non-human entities are of equal legal significance and responsibility must be taken to ensure balance between them (McGregor, 2018, p 19). Justice approaches from Indigenous perspectives would bring relations with environment, nature and all other relatives in Creation into direct conversation with justice discourse in hopes that the epistemic injustice present in dominant discourses of justice can be overcome.

Such a recognition would enable Indigenous approaches to justice to be understood on their own terms. Justice simultaneously can function as a means of recognising and respecting self-determination for Indigenous groups as well as a means towards decolonising relations with dominant societies (Coulthard, 2014; Whyte, 2017). A specifically Indigenous approach to justice, then, may offer insights into how cultural relations of Indigenous peoples emanate from landscape and place (Hernandez, 2019), and point to solutions overcoming the ways in which environmental destruction through 'environmental colonialism' enables the loss of autonomy for Indigenous peoples through the erasure of cultural heritage (Figueroa, 2011). In regard to environmental injustices, Indigenous knowledges, as Whyte (2018, p 70) notes, are not merely supplemental to traditional scientific knowledge, Indigenous knowledges are 'collective capacities that can provide trustworthy and useful wisdom for planning that supports collective self-determination in the face of change'.

Indigenous justice therefore is not one unitary approach but a collection of ontological and epistemological traditions which are unique to each group. For many Indigenous scholars, justice is being advanced when Indigenous peoples have rights of self-determination not only recognised but responded to through formal relations of non-Indigenous communities. Such rights are necessary because they recognise Indigenous knowledges as central for bringing forth healing and balance into all human and non-human relations. Self-determination not only protects but promotes Indigenous legal, social, ontological and epistemological approaches to understanding justice. As such, Indigenous approaches to justice may expand, rather than detract, from substantive framings of justice often siloed in western-dominated ways of thinking.

Notes

[1] These three themes do not exhaust the sentiments of all rights-types expressed in UNDRIP. The right to non-discrimination, for example, is expressed in different articles. But non-discrimination is not unique to Indigenous rights but those of all minority groups more broadly which are also captured in the discourse on 'human rights'.

[2] We acknowledge there are categorial distinctions between the terms 'group' and 'collective' rights. For the purposes herein, the two are used interchangeably.

[3] The UN makes more nuanced distinctions between minority and Indigenous groups than simply assimilation and autonomy-seeking. Kymlicka (2008, p 4) notes that the UN draws out three basic differences between them: 'a) minorities seek institutional

integration while indigenous peoples seek to preserve a degree of institutional separateness; b) minorities seek to exercise individual rights while indigenous peoples seek to exercise collective rights; c) minorities seek non-discrimination while indigenous peoples seek self-government'.

References

Acosta, A. and Martinez Abarca, M. (2018) Buen vivir: An alternative perspective from the peoples of the global south to the crisis of capitalist modernity. In V. Satgar (ed), *The Climate Crisis: South African and Global Democratic Eco-Socialist Alternatives*. Johannesburg: Wits University Press, pp 131–147.

Ananya, S. (1999) Superpower attitudes toward indigenous peoples and group rights. *American Society of International Law Proceedings*, 93, 251–206.

Caillon, S., Cullman, G., Verschuuren, B. and Sterling, E.J. (2017) Moving beyond the human-nature dichotomy through biocultural approaches: Including ecological well-being in resilience indicators. *Ecology and Society*, 22(4), 27.

Capeheart, L. and Milovanovic, D. (2007) Indigenous/postcolonial forms of justice. In *Social Justice*. Ithaca, NY: Rutgers University Press, pp 108–124.

Coulthard, G. (2014) *Red Skin, White Masks: Rejecting the Colonial Politics of Recognition*. Minneapolis, MN: University of Minnesota Press.

Cornell, S. (2018) Justice as position, justice as practice: Indigenous governance at the boundary. In J. Hendry, M. Tatum, M. Jorgensen and D. Howard-Wagner (eds), *Indigenous Justice: New Tools, Approaches, and Spaces*. Bristol, UK: Palgrave, pp 11–26.

Figueroa, R.M. (2011) Indigenous peoples and cultural losses. In J. Dryzek, R. Norgaard and D. Schlosberg (eds), *The Oxford Handbook of Climate Change and Society*. Oxford, UK: Oxford University Press, pp 232–249.

Foley, D. (2003) Indigenous epistemology and standpoint theory. *Social Alternatives*, 22(1), 44–52.

Hernandez, J. (2019) Indigenizing environmental justice: Case studies from the Pacific Northwest. *Environmental Justice*, 12(4), 175–181.

International Labour Organization (2017) C169 – Indigenous and Tribal Peoples Convention, 1989 (No. 169). https://www.ilo.org/dyn/norm lex/en/f?p=NORMLEXPUB:12100:0::NO::P12100_ILO_CODE:C169

Ivison, D. (2003) The logic of aboriginal rights. *Ethnicities*, 3(3), 321–344.

Ivison, D., Patton, P. and Sanders, W. (2000) Introduction. In D. Ivison, P. Patton and W. Sanders (eds), *Political Theory and the Rights of Indigenous Peoples*. Cambridge, UK: Cambridge University Press, pp 1–21.

Kim, E.J.A., Asghar, A. and Jordan, S. (2017) A critical review of traditional ecological knowledge (TEK) in science education. *Canadian Journal of Science, Mathematics, and Technology Education*, 17(2), 258–270.

Kuppe, R. (2009) The three dimensions of the rights of Indigenous Peoples. *International Community Law Review*, 11(1), 103–118.

Kymlicka, W. (1995) *Multicultural Citizenship: A Liberal Theory of Minority Rights*. Oxford, UK: Clarendon Press.

Kymlicka, W. (2008) The internationalization of minority rights. *International Journal of Constitutional Law*, 6(1), 1–32.

Le Grange, L. and Mika, C. (2018) What is indigenous philosophy and what are its implications for education. In P. Smeyers (ed), *International Handbook of Philosophy of Education*. Cham, Switzerland: Springer International Publishing, pp 499–515.

McGregor, D. (2018) Achieving Indigenous environmental justice in Canada. *Environment and Society: Advances in Research*, 9(1), 7–24.

Newman, D.G. (2007) Theorizing collective indigenous rights. *American Indian Law Review*, 31(2), 273–290.

Patton, P. (2019) Philosophical foundations for Indigenous economic and political rights. *International Journal of Social Economics*, 46(11), 1264–1276.

Pierotti, R. (2010) *Indigenous Knowledge, Ecology, and Evolutionary Biology*. London: Routledge.

Sanderson, D. (2012) Redressing the right wrong: The argument from corrective justice. *University of Toronto Law Journal*, 62(1), 93–132.

Sarivaara, E., Maata, K. and Uusiautti, S. (2014) Who is indigenous? Definitions of indigenity. *European Scientific Journal*, 9(10), 369-378.

Scheinin, M. (2005) What are indigenous people? In N. Ghanea-Hercock and A. Xanthaki (eds), *Minorities, People, and Self-Determination: Essays in Honor of Patrick Thornberry*. Leiden, The Netherlands: Brill, pp 3–13.

United Nations (2008) *United Nations Declaration on the Rights of Indigenous People*. https://www.un.org/development/desa/indigenouspeoples/declaration-on-the-rights-of-indigenous-peoples.html

UNPFII (United Nations Permanent Forum for Indigenous Issues) (2007) Who are indigenous? https://www.un.org/esa/socdev/unpfii/documents/5session_factsheet1.pdf

Valadez, J.M. (2012) Indigenous rights. In D. Callahan and P. Singer (eds), *Encyclopedia of Applied Ethics*. London: Academic Press, Elsevier, pp 696–703.

Vieille, S. (2012) Māori customary law: A relational approach to justice. *International Indigenous Policy Journal*, 3(1), Article 4, 1–18.

Watene, K. (2016a) Indigenous peoples and justice. In K. Watene and J. Drydyk (eds), *Theorizing Justice: Critical Insights and Future Directions*. London: Rowman & Littlefield, pp 133–152.

Watene, K. (2016b) Valuing nature: Māori philosophy and the capability approach. *Oxford Development Studies*, 44(3), 287–296.

Watene, K. (2020) Transforming global justice theorizing: Indigenous philosophies. In T. Brooks (ed), *The Oxford Handbook of Global Justice.* Oxford: Oxford University Press, pp 163–180.

Weigård, J. (2008) Is there a special justification for indigenous rights? In H. Minde (ed), *Indigenous Peoples: Self-Determination, Knowledge, Indigenity.* Delft, The Netherlands: Eburon Academic Publishers, pp 177–192.

Whyte, K. (2017) Indigenous climate change studies: Indigenizing futures, decolonizing the Anthropocene. *English Language Notes,* 55(1–2), 153–162.

Whyte, K. (2018) What do indigenous knowledges do for indigenous peoples? In M. Nelson and D. Shilling (eds), *Traditional Ecological Knowledge: Learning from Indigenous Practices for Environmental Sustainability.* Cambridge, UK: Cambridge University Press, pp 57–82.

Widenhorn, S. (2014) Towards epistemic justice with indigenous peoples' knowledge? Exploring the potentials of the convention on biological diversity and the philosophy of buen vivir. *Development,* 56(3), 378–386.

8

The Capabilities Approach

Stephen Przybylinski and Roman Sidortsov

Introduction

The Capabilities Approach (CA) is not a normative theory of justice as it is a method used for evaluating human well-being. With that said, the CA has been used extensively to compare and measure equality among individuals within given societies. This directly connects the approach with certain conceptions of justice. The CA can generally be defined as an evaluative framework which assesses how well people are able to realise certain opportunities enabling them to lead a dignified life. It has been utilised to assess whether people are living below a certain threshold which inhibits them from obtaining basic resources or securities, for instance, and therefore addresses whether they have the opportunity or capability to lead a healthy or dignified life. Although not theorising justice explicitly, the approach emerged from critiques of justice theorising. This chapter therefore provides an overview of the development of the concept, primarily focusing on the analytical development of the approach, as it relates to justice.

The terms 'functionings' and 'capabilities' are central concepts to the approach. The terms underscore the normative value CA theorists place upon enabling individuals the opportunities to lead more dignified lives. Amartya Sen, who developed the original notion of capabilities, states that living can be defined as 'a set of interrelated "functionings" consisting of beings and doings. A person's achievement in this respect can be seen as the vector of his or her functionings' (Sen, 1992, p 39). Here, 'functionings' are understood as being well nourished, receiving an education or being able to travel. Closely related to the notion of functionings is that of having the *capability* to function. For CA theorists, having the capability to function means that there are 'various combinations of functionings (beings and doings) that the person can achieve ... reflecting the person's freedom to lead one type of life or another' (Sen, 1992, p 40). Capabilities are the 'real' or substantive freedoms people have to achieve certain functions. For CA scholars, then, evaluating well-being is to measure people's 'capability to function', or the 'effective opportunities' that individuals have 'to undertake the actions and activities that they want to engage in, and be whom they want to be' (Robeyns, 2005, p 95). Evaluating the capability to function is dependent on how any given study defines a function.

Given its focus on measuring well-being, the CA is used by its interlocutors as a method for assessing it. Perhaps because of this, most CA theorists argue against the need to establish CA as an ideal theory of justice. In other words, the CA does not attempt to analytically construct a just 'basic structure of society' like that found in the liberal egalitarian theories of Rawls and Dworkin (see Chapter 1). Although Sen (2009) acknowledges that the conceiving of ideal institutions is one step towards enabling more equitable societies, he argues that what is needed to address social inequality is an assessment of actual human well-being in connection with principles of 'just' institutions. As such, Sen advocates a 'realization-based comparison' approach for assessing equality which evaluates 'the advancement or retreat of justice' (2009, p 8). A comparative or 'realization-focused perspective', he argues, 'makes it easier to understand the importance of the prevention of manifest injustice in the world, rather than seeking [to identify] the perfectly just' (2009, p 21). For Sen, the CA aims not at identifying ideal and just basic social structures but one that compares and evaluates what is unjust within 'actually-existing' social relations.

With a focus on measuring well-being in actually-existing societies, the approach has been used quite broadly. A range of academic disciplines utilise the method for framing their studies, most notably within development and poverty studies (Robeyns, 2006), but also within the fields of health (Mitchell et al, 2017), education (Walker, 2005), gender studies (Robeyns, 2003) and environment–society relations (Holland, 2008). Governmental agencies have also adopted the framework. For instance, the capabilities language has been adopted to guide policy making in the *Human Development Report* of the United Nations (UNDP, 2020). Despite the widespread use of the CA as a method, however, its application to justice theorising has been a source of debate. As such, the following sections dive further into prominent critiques of the CA, which help illustrate the ways in which the approach does and does not engage with the concept of justice. Following these critiques, we end the chapter by identifying the potential uses of the approach to justice theories.

Substantive and procedural approaches to capabilities theory

If Sen provided the initial critique of ideal theories of justice that lead to the CA in general, it was Martha Nussbaum who developed the approach's principles closer to a theory of justice. More so than Sen, Nussbaum has identified certain moral criteria that define key types of capabilities for an equitable society upon which individuals might agree. For instance, Nussbaum (2000; 2003; 2011) has refined a list of ten 'central capabilities' which establishes a bare minimum or threshold for the requirements of a 'dignified life'. The list includes the right to live a normal length life, bodily integrity and health, among others. Her basic moral claim is that 'respect for human dignity requires that citizens be placed above an ample (specified) threshold of capability, in all ten of those areas' (Nussbaum, 2011, p 36). For Nussbaum, the capabilities on this list could be directly incorporated as guarantees into a state constitution, such as the Indian Constitution, which she often refers to in her work.

Nussbaum's political goal in working towards such a capability standard is that:

> [A]ll should get above a certain threshold level of combined capability, in the sense not of coerced functioning but of substantial freedom to choose and act. That is what it means to treat all people with equal respect. Accordingly, the attitude toward people's basic capabilities is not a meritocratic one – more innately skilled people get better treatment – but, if anything, the opposite: those who need more help to get above the threshold get more help. (Nussbaum, 2011, p 24)

The point for Nussbaum is to provide an objective standard by which to evaluate whether individuals do or do not meet the threshold of these ten capabilities.

Nussbaum stops short of developing a normative theory of justice, however. 'The capabilities approach is not a theory of what human nature is, and it does not read norms off from innate human nature', she states. 'Instead, it is evaluative and ethical from the start: it asks, among the many things that human beings might develop the capacity to do, which ones are the really valuable ones, which are the ones that a minimally just society will endeavour to nurture and support?' (Nussbaum, 2011, p 28). Similar to Sen, therefore, Nussbaum sees the CA as a comparative quality-of-life assessment, one 'taking each individual as an end, and asking not just about the total or average well-being but the opportunities available to each person' (2011, p 18). What is notable about Nussbaum's political conception of the CA is that it values a pluralist notion of citizens being free to choose how they realise their 'functionings' if and when they have these central capabilities protected. Nussbaum sees this flexibility as being more culturally sensitive than in Rawls' notion of justice grounded in distributive goods, as it stresses that gender and racial inequities, for example, require culturally specific attention.

Although the CA does not advance a normative theory of justice, the approach has been developed by some scholars as a procedure through which inequality may be evaluated, connecting it with a basic concept of non-ideal justice. Ingrid Robeyns, for instance, argues that 'if we want to respect Sen's capability approach as a general framework for normative assessments, then we cannot endorse one definite list of capabilities without narrowing the capability approach' (Robeyns, 2006, p 69). At the same time, Robeyns notes that if the CA is to be of use, some selection of criteria must be made regarding *which* capabilities are necessary for comparative purposes. Rather than create an objective list or constitution of central capabilities, Robeyns argues instead for establishing a procedure through which scholars may select key capabilities most relevant to a given study.

The procedure Robeyns puts forward argues simply that there ought to be certain principles used to select a given study's capabilities and *not* what capabilities must be included in *all* studies of well-being. Robeyns (2006, pp 70–71) identifies five criteria necessary for selecting 'functions' through which to assess capabilities. The first is *the criterion of explicit formulation* which states that any chosen function must be explicit to all, necessitating a process of discussion and defence of each function. The second is the criterion of *methodological justification* which states that scholars must 'clarify and scrutinize the method that has generated the list and justify this as appropriate for the issue at hand'. Third is the criterion of *sensitivity to context* which suggests that the level of abstraction of the list of functions must be relevant to the capabilities being evaluated. Depending on which functions make the list, researchers ought to remain sensitive to how a chosen function speaks to the situations surrounding the study. The fourth criterion states there should be *differing levels of generality* when choosing functions. When selecting functions, researchers should choose both 'ideal' functions as well as 'second best' functions to remain flexible to the range of experiences acknowledged during the research process. The final criterion is *exhaustion and non-reduction*, or that all listed functions exhaust 'all important elements' which should 'not be reducible to other elements'. With these selection criteria, Robyens offers the CA a procedure that avoids narrowing down 'essential' capabilities while also affirming that the approach can have a replicable procedure by which to evaluate well-being at the level of the non-ideal.

If there is one core capability that all capability theories share, Robeyns (2016) argues, it is that 'functionings and capabilities' must be the foundation for any theory within this familial approach. 'From this core there then can be a variety of modules that detail the purpose of a given capability theory, its meta-ethical commitments, empirics, and additional normative commitments' (Robeyns, 2016, pp 403–404). Seen not as ideal theory but as a procedure or methodical framing, then, CA scholars note the existence of a range of situations through which to evaluate inequalities. However, as a means of identifying what constitutes equity or justice, CA scholars remain uncertain. The following section addresses some of the key debates throughout the development of the CA which illustrate its connection to justice theories.

Debates and key controversies

What type of equality?

The CA emerged as a critique of distributive (resource) theories of justice in Amartya Sen's early work. Sen has long been critical of resource based theories of justice, such as those provided by Rawls, Dworkin (Chapter 1) or Nozick (Chapter 2), questioning in particular what type of equality such theories of justice are attempting to achieve. Rawls, for instance, seeks equal opportunities for individuals; Dworkin seeks equality of resources; while Nozick seeks equal protection of individual liberty rights. Sen argues these distributive theories do not give an accurate sense of how equal individuals actually are because they emphasise and measure one aspect of equality: the distribution of resources or primary goods (income or rights). Given this focus, Sen argues that distributive theories do not get at either how 'free' individuals actually are or what capabilities individuals have to pursue well-being. Equality is better understood, he argues, through the 'actual freedom that is represented by the person's "capability" to achieve various alternative combinations of functionings' (Sen, 1992, p 81). For Sen, capability 'represents *freedom*, whereas [the distributions of] primary goods tell us only about the *means* to freedom, with an interpersonally variable relation between the means and the actual freedom to achieve' (1992, p 84, emphasis in original). To avoid the narrowness of assessing justice solely through the distribution of resources, Sen argues we need to 'examine interpersonal variations in the transformation of primary goods ... into respective capabilities to pursue our ends and objectives' (1992, p 87). That is, we may assess equality by how individuals are able to use resources to 'realise' certain capabilities.

The CA was initially defined in part by breaking with ideal resource theories of justice by urging that equality is better understood comparatively, by evaluating individuals' ability to realise dignified lives via their own ends. Indeed, the CA has had wide appeal because of its flexible method of evaluation. Yet, it is exactly the lack of well-developed theoretical criteria that has led some to criticise the usefulness of the approach. Thomas Pogge (2002; 2010), for instance, argues that the CA does not establish a publicly derived standard of justice, such as that found in Rawls' justice as fairness (see Chapter 1), and as such, it does not offer a way to compare whether individuals are treated unequally. The notion of a capability theory of justice is further troubled by the open-ended definition of a capability. As Richard

Arneson (2010) notes, if we are even able to identify what a capability is in the first place, by nature, individuals hold incommensurable capabilities in relation to one another. Such critiques identify the difficulty in evaluating how one individual's ability to realise 'functions' in relation to another individual makes comparing well-being more difficult and relative than in ideal resource theories of justice.

Individualism

Related to debates over resource theories of, and capability approaches to, justice is the notion that the CA is too individualistic in scope. Above all, the CA emphasises that equality is found when individuals are free to choose the things they value in realising their well-being. Given the focus on individual well-being, some of the early critiques of the CA suggest that the approach is unable to connect well-being beyond the scale of the individual. As Hartley Dean (2009) argues, the CA's metric for evaluation compares individuals in abstraction; 'the person being compared and "the other" interact in a metaphorical "space of capabilities"', whereby 'both the person and the other are constituted as the abstract bearers of capabilities' (p 268). In other words, the approach assumes that individuals have capabilities which are disconnected from those of other individuals.

Many have argued that the 'methodological individualism' upon which CA is premised, therefore, does not acknowledge the societal relations on which individuals depend to realise their 'capabilities' and 'functionings'. The individual becomes the subject of injustice. 'The priority is individual liberty, not social solidarity; the freedom to choose, not the need to belong. In the space of capabilities the individual is one step removed; she is objectively distanced from the relations of power within which her identity and her life chances must be constituted' (Dean, 2009, p 267). What takes precedence in CA theories, Charles Gore (1997, pp 243–244) argues, 'is not simply the nature of individual lives ... but [that] the objects of value are also what individuals "have". That is to say, the objects of value are "properties" of individuals, in the sense that they belong to them'. As such, the CA 'requires that judgments about the goodness of states of affairs are based exclusively on properties of individuals' (Gore, 1997, p 243). Prioritising individuals' values of the good life may further connect a theory of well-being more generally.

Claims of being too individualistic are thus brought from a more communitarian perspective. Communitarians see in the CA that 'objects of value' with which individuals choose to realise their well-being negate an evaluation of the societal properties which also contribute to an individual's well-being. One effect of this narrowness, Gore (1997, p 247) argues, is that the CA can give 'a flawed view of inequalities in individual well-being in culturally heterogeneous societies, in cross cultural comparisons of well-being and where functionings are not so basic that they are simply biological' (see also Chapter 7 for an Indigenous critique of the individualism of the CA). Social context is necessary, as Stewart and Deneulin (2002, p 67) note, as 'both the extent of agency and the objectives that people value depend in part on the environment in which the individual lives. Hence one needs to assess the structures which influence agency and the formation of objectives'. The charge of individualism resonates with other perspectives within the CA, like that of Nussbaum, who argues for a definitive set of capabilities in order to establish objective values for capabilities.

The critique of CA as being centred exclusively on an individual has led to the emergence of the concept of collective capabilities (Evans, 2002). Under the conventional CA, the capability to function is attached to an individual, thereby making opportunities for collective entities, local communities, for example, to undertake collective actions and activities that they want, an aggregate of individual capabilities of members of such entities. However, this all but rejects the collective process of sharing, debating and learning through one another's experiences about what is valued and how best to act on the valued outcomes. A discourse, which is inherently collective, allows people to engage in deliberation and debate on their common concerns and experiences. The ability to have such a discourse and formulate a common capability to function is not a sum of individual capabilities. Rather, it is a function of a collective that co-exists and complements individual capabilities. According to Evans (2002), the access and opportunity of individuals 'to do things that they have a reason to value' is only possible via *collective* decisions about distribution of benefits and impacts among different members of a collective entity. These collective decisions lead to acquiring and formulating individual capabilities.

Defining 'central' capabilities

Whether or not a definitive list of capabilities should or can be established is a primary tension in the capabilities literature. Establishing a capability theory of justice would require that some principles of justice be established. Nussbaum's list of central capabilities can be seen as a beginning of a justice theory based on capabilities. Yet, few accept Nussbaum's push for a capability theory of justice, preferring instead to use the approach as a general method for framing studies of well-being, however defined. Most notably, Sen (2005) has abstained from endorsing an objective list of central capabilities. Like Sen, Robeyns sees the strength of the approach being its flexibility, an evaluative tool which does not need to develop principles of justice in order to hold utility for addressing inequalities. Yet, Robeyns (2003) notes there are drawbacks to turning away from theorising capabilities as a conception of justice. For instance, the wide application of the CA framework brings about quite divergent normative results, as those using the approach draw from a great variety of social theories. Although most using CA appreciate its flexibility to incorporate different social theories into the evaluative framework, this can lead to a loosely defined understanding of what constitutes the capability to function as there are no established or central capabilities.

If a normative CA theory of justice was to be pursued, such a theory, Arneson (2010, p 108) argues, would require 'a standard that distinguishes significant from trivial capabilities and discounts the latter in the comparison of people's condition in terms of their capability sets'. A capability theory of justice would not weigh all capabilities equally, for valuing each individual capability equally would make pointing to injustice or inequity next to impossible. 'The capabilities that matter for purposes of the theory of justice are capabilities to achieve or be what is objectively good, what contributes to the quality of one's life as rated by an objective list account of human good' (Arneson, 2010, p 108). As such:

> [T]he basis of interpersonal comparison for the theory of justice is best regarded as capability to live a life that is objectively worthwhile – not merely what the individual subjectively regards as such. The capabilities of an individual that matter for social justice obligations are capabilities as ranked and ordered by an objective list or perfectionist conception of well-being. (Arneson, 2010, p 122)

Arneson's arguments underscore that any theory of justice must incorporate a moral distinction of what constitutes not *a* good, but some sense of what is the *collective* good. As Martijn Boot (2012, p 8) argues, a capabilities based theory of justice need not identify perfect or ideal justice. Rather, the question is whether, 'to be capable of adequately comparing different social states with respect to degrees of justice, we need to identify and order criteria of justice that can serve as standards of comparison'. Like 'perfect' theories of justice that Sen rejects, there must, Boot (2012, p 8) argues, be *some* integrated principles of justice 'to be able to adequately compare imperfect social states'. As previously mentioned, few CA theorists agree on a set of principles of justice based on generally applicable capabilities.

Underemphasising political economy and colonialism

A few critiques of the CA centre on the relative absence of the role that economic relations play in the approach. For instance, Pogge (2002, p 217) argues that metrics used by capability theorists such as Nussbaum 'tend to conceal the enormous and still-rising economic inequalities which resource metrics (Rawlsians) make quite blatant'. The result of this is that CA may 'exaggerate the relative aspects of poverty', thereby downplaying the utility of political economic critique in highlighting certain situations of oppression which are related to resource allocation. Following this, others have argued that the relative focus on individual oppressions misses an opportunity to challenge how the ordering of the global economy structures such inequalities in the first place. Critiquing Nussbaum's list of capabilities, Jaggar (2002) argues that such a universal list misses the ways in which global economic relationships structure local conditions of gender oppression, for example.

Further, Jaggar (2002) argues that western philosophers should scrutinise the role of western countries in exacerbating oppressions related to gender, race and class instead of solely focusing on injustices in non-western states. 'Instead of developing new standards ... to scrutinize Nonwestern practices', she argues, 'the priority for Western philosophers should be to examine critically the ways in which our own countries are implicated in impoverishment and political marginalization of women in the global South' (Jaggar, 2002 p 230). Jaggar's call attempts to avoid colonising inequalities, as she charges Nussbaum with doing, so as to better critique the structures establishing iniquitous conditions for people. That, Jaggar suggests, the CA has not shown itself to be able to do just yet.

As the previous two sections illustrate, the CA is not a normative theory of justice as much as it is a method by which to evaluate individual well-being. The CA thus helps develop insights about inequality rather than directly developing principles to establish a theory of social justice. Whether the CA ought to develop a theory of justice is beyond the scope of this overview. What can be said about what the CA offers researchers interested in justice is that it provides a method with which to make connections to social inequality and injustice more generally. To make connections with the justice literature, researchers must be explicit about how they identify which functions and capabilities establish more equitable social relations, not just individual well-being. That is, researchers would need to justify how their chosen capabilities or functions move beyond the relative freedoms of the individual to express a general state of unjust relations more broadly. The final section addresses what the CA may offer social science researchers examining well-being as a proxy for injustice.

Applying the Capabilities Approach in the social sciences

The CA is widely used by scholars and policy makers alike and provides an evaluative framework through which to assess matters of inequality in a variety of situations. As Robeyns (2006, pp 360–361) notes, the CA has been used as an evaluative tool for:

- general assessments of the human development of a country;
- the assessment of small-scale development projects;

- identification of the poor in developing countries;
- poverty and well-being assessments in advanced economies;
- an analysis of deprivation of disabled people;
- the assessment of gender inequalities;
- theoretical and empirical analyses of policies;
- critiques of social norms, practices and discourses; and
- the use of functionings and capabilities as concepts in non-normative research.

While such a range illustrates how inequality can be approached when using a CA framework, it will be up to researchers to justify how the CA method advances concepts of justice.

As a method for evaluating inequality, the CA may be used in combination with different theories of justice. The CA's identification with political liberalism, while sensitive to gender and racial inequalities, makes this approach a very adaptable and generalised metric through which to assess well-being. In what follows, we briefly point to connections between two 'applied' justice theories, environmental and energy justice (see Chapters 9 and 11), to illustrate how the CA contributes to the development of these justice approaches.

Environmental justice

Holland (2008) notes that neither Nussbaum nor Sen recognise a particular importance of the environment for developing the CA. She sees that as a shortcoming of both scholars' work and makes a case for acknowledging the instrumental value of the environment for human capabilities. She further argues that the environment should be considered as an independent 'meta-capability' manifesting certain environmental conditions. The instrumentality of the environment for human capabilities is a particularly strong point because it establishes a foundation upon which capabilities can be recognised. For example, the environment (and by extension clean water and air and healthy ecosystems among others) appear to be necessary conditions for Nussbaum's central capabilities such as life, bodily health and other species.

Schlosberg and Carruthers (2010) make this point explicit by showing how environmental problems experienced by Indigenous peoples impact the capabilities of communities and their individual members. They do so through two case studies of Indigenous communities in the United States and Chile. Ballet et al (2013) argue that the CA is essential for defining sustainable development. They note that: 'taking the freedom to achieve into account radically alters the assessment of well-being, making it possible to establish a direct link with environmental sustainability, with regard to the extent to which freedom is transmissible within and between generations' (Ballet et al, 2013, p 29). They adopt Martins' argument that because future generations might value outcomes differently, imposing outcomes based on the values of the present generation is deficient because it circumvents their freedom of choice. Therefore, sustainable development should be premised on the sustainability of freedom of choice.

Therefore, collective capabilities provide a particularly useful lens for addressing environmental justice issues faced by communities. Whether it is largely African American communities living nearby refineries on the Gulf of Mexico coast or Indigenous communities struggling with oil field pollution in Siberia, unfair distribution of environmental impacts and risks affects the ability of collective entities to flourish. Virtually nonexistent tax revenue and poor public education are two interrelated consequences of unfair distribution of pollution that diminish the collective capabilities of these communities to flourish.

Energy justice

Energy justice is a novel and rapidly developing concept, which is yet to receive the same level of cross pollination with normative theories of justice as environmental justice. One of the exceptions is the CA; due to its flexible and applied nature, the CA has been embraced by energy scholars. For example, Sovacool and Dworkin argue (2014, p 437) that energy poverty 'interferes with human beings' ability to achieve functions and capabilities'. Jenkins et al (2016) note the impact of energy development on the capabilities of Indigenous communities. Hillerbrand et al (2021) explore whether Nussbaum's central capabilities can be used to construct an analytical framework for the assessment of digitisation of the energy sector. Baard and Melin (2022) develop the limits, thresholds and ceilings of energy justice based on the CA.

The work of Jones et al (2015) represents one of the earliest scholarly attempts to connect energy justice and the CA. They do so by constructing four foundational assumptions en route to defining energy justice. These four assumptions are as follows:

1. Every human being is entitled to the minimum of basic goods of life that is still consistent with respect for human dignity.
2. The basic goods to which every person is entitled also include the opportunity to develop the characteristically human capacities needed for a flourishing human life.
3. Energy is only an instrumental good – it is not an end in itself.
4. Energy is a material prerequisite for many of the basic goods to which people are entitled. (Jones et al, 2015, pp 151–160)

The result of Jones et al's (2015) conceptualisation is the affirmative and prohibitive principles of energy justice:

- Energy systems must be designed and constructed in such a way that they do not unduly interfere with the ability of any person to acquire those basic goods to which he or she is justly entitled.
- If any of the basic goods to which every person is justly entitled can only be secured by means of energy services, then in that case there is also a derivative right to the energy service.

It is important to note that although Jones and his colleagues do not use the term 'capabilities', they premise assumptions two and four on the notion of capabilities. In the second assumption, they adopt Nussbaum's and Sen's rejection of the finality of basic goods. In the fourth assumption, they make a similar argument to that of Holland recognising the instrumentality of energy for developing human capabilities.

Tarekegne and Sidortsov (2021) draw from the guiding principles of recognition, distribution, procedure, and restoration in the context of electricity access in sub-Saharan Africa by merging the Affirmative Principle (AP) and Prohibitive Principle (PP) and the CA. However, they do stop short of reforming the principles as based on capabilities; nor do they elaborate on the individual and collective capabilities as part of the CA. Sidortsov and Badyina (2023) remedy this issue by restating the principles:

- PP: energy systems must be designed and constructed in such a way that they do not unduly interfere with individual and/or collective capabilities.
- AP: energy services must be provided if they are instrumental to securing individual and/or collective capabilities.

Conclusion

Although the CA has not developed a normative theory of justice, its value for studying and conceptualising justice is clear. The focus on evaluating human and non-human well-being as a metric of inequality directly relates the CA to this central concept of justice. As the chapter has shown, the means by which CA scholars assess inequality differ. Some focusing on human development look at how levels of poverty prevent individuals and communities from having the capability to flourish without enjoying basic resources. Others assess the ways in which environmental conditions affect human and non-human well-being, or how energy poverty or gender inequity prevents individuals from having the opportunity to lead dignified lives. The potential application of the CA as a methodology used to make transparent the iniquitous outcomes of various processes is wide-ranging.

That researchers continue to utilise the CA in relation to the concept of justice is important. Its insights can be useful for identifying inequities that can be empirically and normatively analysed. The contextuality of the CA is also important because it enables researchers to utilise the CA in conjunction with applied theories of justice. To ensure that the freedom or capability to achieve well-being remains in concert with conceptions of justice, the CA must not only compare indexes of inequity, but identify how given inequities are unjust and what ought to be done to make them just. Not all CA theorists see more normative theorising of justice as the objective of this approach. But theorising justice remains a fruitful pursuit for CA theorists seeking to evaluate how well-being may be developed into an account of justice, an account of justice that shows us what is required for individuals who are below a given threshold of well-being to overcome that situation. This is not to say that the CA is not normatively inclined to describe how well-being may be better realised for those lacking the capability or opportunity to 'flourish'. In this sense, and to its benefit, the CA remains open as an approach, available to theorists interested in justice theorising or for those looking to examine indexes of well-being that help us identify a variety of metrics for assessing the problems resulting from inequalities.

References

Arneson, R.J. (2010) Two cheers for capabilities. In H. Brighouse and I. Robeyns (eds), *Measuring Justice*. Cambridge, UK: Cambridge University Press, pp 101–128.

Baard, P. and Melin, A. (2022) Max power: Implementing the capabilities approach to identify thresholds and ceilings in energy justice. *Science and Engineering Ethics*, 28(1), 8.

Ballet, J., Koffi, J.M. and Pelenc, J. (2012) Environment, justice, and the capability approach. *Ecological Economics*, 85(2013), 28–34.

Boot, M. (2012) The aim of a theory of justice. *Ethical Theory and Moral Practice*, 15(1), 7–21.

Dean, H. (2009) Critiquing capabilities: The distractions of a beguiling concept. *Critical Social Policy*, 29(2), 261–278.

Evans, P. (2002) Collective capabilities, culture, and Amartya Sen's *Development as Freedom*. *Studies in Comparative International Development*, 37(2), 54–60.

Gore, C. (1997) Irreducibly social goods and the informational basis of Amartya Sen's capability approach. *Journal of International Development*, 9(2), 235–250.

Hillerbrand, R., Milchram, C. and Schippl, J. (2021) Using the capability approach as a normative perspective on energy justice: Insights from two case studies on digitalisation in the energy sector. *Journal of Human Development and Capabilities*, 22(2), 336–359.

Holland, B. (2008) Justice and the environment in Nussbaum's 'capabilities approach': Why sustainable ecological capacity is a meta-capability. *Political Research Quarterly*, 61(2), 319–332.

Jaggar, A.M. (2002) Challenging women's global inequalities: Some priorities for western philosophers. *Philosophical Topics*, 30(2), 229–252.

Jenkins, K., McCauley, D., Heffron, R., Stephan, H. and Rehner, R. (2016) Energy justice: A conceptual review. *Energy Research & Social Science*, 11, 174–182.

Jones, B., Sovacool, B. and Sidortsov, R. (2015) Making the ethical and philosophical case for 'energy justice'. *Environmental Ethics*, 37(2), 145–168.

Mitchell, P.M., Roberts, T.E., Barton, P.M. and Coast, J. (2017) Applications of the capability approach in the health field: A literature review. *Social Indicators Research*, 133, 345–371.

Nussbaum, M. (2000) *Women and Human Development: The Capabilities Approach*. Cambridge, UK: Cambridge University Press.

Nussbaum, M. (2003) Capabilities as fundamental entitlements: Sen and social justice. *Feminist Economics*, 9(2–3), 33–59.

Nussbaum, M. (2011) The central capabilities. In *Creating Capabilities: The Human Development Approach*. Cambridge, MA: Harvard University Press, pp 17–45.

Pogge, T. (2002) Can the capability approach be justified? *Philosophical Topics*, 30(2), 167–228.

Pogge, T. (2010) A critique of the capability approach. In H. Brighouse and I. Robeyns (eds), *Measuring Justice*. Cambridge, UK: Cambridge University Press, pp 17–60.

Robeyns, I. (2003) Sen's capability approach and gender inequality: Selecting relevant capabilities. *Feminist Economics*, 9(2–3), 61–92.

Robeyns, I. (2005) The capability approach: A theoretical survey. *Journal of Human Development*, 6(1), 93–117.

Robeyns, I. (2006) The capability approach in practice. *Journal of Political Philosophy*, 14(3), 351–376.

Robeyns, I. (2016) Capabilitarianism. *Journal of Human Development and Capabilities*, 17(3), 397–414.

Schlosberg, D. and Carruthers, D. (2010) Indigenous struggles, environmental justice, and community capabilities. *Global Environmental Politics*, 10(4), 12–35.

Sen, A. (1992) *Inequality Reexamined*. Cambridge, MA: Harvard University Press.

Sen, A. (2005) Human rights and capabilities. *Journal of Human Development*, 6(2), 151–166.

Sen, A. (2009) *The Idea of Justice*. Cambridge, MA: Harvard University Press.

Sidortsov, R. and Badyina, A. (2023) Expanding collective capabilities to conceptualise and assess the impact of oil and gas activities on the energy transition in the Arctic. In C. Wood-Donnelly and J. Ohlsson (eds), *Arctic Justice: Environment, Society & Governance*. Bristol, UK: Bristol University Press. https://www.tandfonline.com/doi/abs/10.1080/1464988050 0120491

Sovacool, B.K. and Dworkin, M.H. (2014) *Global Energy Justice: Problems, Principles, and Practices*. Cambridge, UK: Cambridge University Press.

Stewart, F. and Deneulin, S. (2002) Amartya Sen's contribution to development thinking. *Studies in Comparative International Development*, 37(2), 61–70.

Tarekegne, B. and Sidortsov, R. (2021) Evaluating sub-Saharan Africa's electrification progress: Guiding principles for pro-poor strategies. *Energy Research & Social Science*, 75, 102045.

United Nations Development Programme (UNDP) (2020) *Human Development Report 2020: The Next Frontier: Human Development and the Anthropocene*. New York: UNDP.

Walker, M. (2005) Amartya Sen's capability approach and education. *Educational Action Research*, 13(1), 103–110.

PART II

Applied Justice Theories

As the chapters in Part I collectively show, justice theories are complex, widely ranging in their focus, principles and arguments. Though diverse, what binds them is their collective pursuit to refine how we come to know what is just and unjust. They do so not only by proposing idealised conceptions of what is or is not just, but also by critically deconstructing such idealised theories to indicate, for instance, what non-ideal justice theories help us to see. Such a pursuit has been centuries long in the making. As we saw in Part I, liberal discourses of justice began to substantively develop from arguments made in the 17th century, but that tradition can be linked back to ancient Greece along with the cosmopolitan tradition's roots, which were birthed by the Stoics in ancient Greece. The insights derived from such theorising undoubtedly advance how we make sense of the concept of justice. But these approaches do not exhaust the ways in which justice can be understood. Further approaches have developed more recently, within the last half-century or so, applying these insights within disciplinary or topically specific concerns. These general bodies of justice scholarship are what we call 'applied' theories of justice, which emerge from the fields of justice theorising that are focus of the chapters in Part II.

In this volume, we make the distinction between normative theories (Part I) and applied fields of justice (Part II) because the latter have generally developed with different aims in mind, compared with the broader philosophical and normative discussions of justice. The applied fields of justice are bodies of scholarship intending to work through justice theories specifically in relation to topical fields of practice and research, for example, the relation between environmental inequity and proper procedures for realising justice. What links these applied fields of research is that they aim to understand justice as grounded in a specific object: the environment, the climate, energy systems, geographical space, the landscape, intergenerational features of justice, or 'just transitions'. These discourses are here seen as distinct fields of their own and were selected based on their prominence within socio-environmental research on justice in addition to their social relevance more broadly.

This is not to say that applied fields of justice remain only within interpretive or descriptive levels of research. By using the term 'applied', we do not want to suggest that these bodies of scholarship lack theoretical rigour or that normative theorising does not shape the ways in which more grounded research examines matters of justice. The chapters on applied fields of justice all have developed, or are beginning to develop, substantive connections between a specialised area of social science research and the concept of justice. The spatial justice literature, for instance, has enjoyed decades-long engagement with normative justice theorising. Others, like energy justice, have only relatively recently begun to engage in the justice literatures, thereby expanding these discourses beyond energy security into the language of justice and inequity. The chapters, then, detail how these bodies of scholarship emerged from research on specific topical issues, highlighting the ways in which justice research can be fruitfully analysed through inquiries foundational to the social and political sciences.

Just as the chapters in Part I provided overviews of the main concepts and debates in specific traditions or schools of thought, so too do the chapters in Part II by expanding and exploring the ways in which we approach and research matters of justice and injustice. The topics for the chapters were chosen due to their prominence within social and political science scholarship on justice, considered both for their historical significance complementing social justice movements as well as their distinction in advancing the social science scholarship itself. While many theories or traditions of justice exist outside of the chapters in Part II, selection necessitates making choices. We believe the *topics* included in Part II chapters will not be unfamiliar to many readers, though we believe that the authors' arguments will reflect these justice traditions with new insight.

Emphasis similarly remains on highlighting the collective development of each field. While the format of the following chapters remains relatively the same as those in the previous part, most of the following chapters also analyse the connection or absence of normative justice theorising within a given field of scholarship. We encourage readers to look for connections and departures that the applied justice traditions make with the more philosophical and normative justice theories represented in Part I.

Stephen Przybylinski and Johanna Ohlsson

9

Environmental Justice

Corine Wood-Donnelly

Introduction

The contextual background to justice theory labelled as environmental justice has a strong tradition in social justice movements related to race and environmental inequality in the United States. However, there are other strands that weave themselves into the thinking on justice theory situated within environmental justice. This includes earlier traditions that are rooted in what is now known as ecological justice, which tidily couples with environmental ethics and resource conservation as well as an older European tradition concerned with the health of urban populations and, more recently, global considerations of environmental justice in the context of international environmental concerns, such as climate change or transboundary pollution, for example (Kuehn, 2000; Taylor, 2000; Schlosberg, 2007; Villa et al, 2020). It is notable that the environment in the mainstream literature on environmental justice uses the term environment as an expression of one's surroundings over the use of the environment as part of nature and, as a result, the mainstream literature is centred on the human experience of their environment over justice for nature itself.

It is common within the environmental justice literature to focus on the forms of distribution, procedure, retribution and recognition across a range of vectors, including race, class, ethnicity, gender and disability, among others. The literature included within environmental justice is vast, situated within a range of disciplines, including sociology, law, geography, political ecology, and others. Much of this scholarship includes empirical and substantive studies of topics related to exposure and proximity to negative environmental indicators such as air pollution, toxic spills, noise, homeownership, water quality and energy poverty or access to positive environmental amenities such as parks, green infrastructure and other green spaces or benefits from nature. However, there have also been efforts to theorise environmental justice (Schlosberg, 2007; 2013) or to frame a taxonomy for understanding the approach (Kuehn, 2000).

Environmental justice is defined through various expressions that attempt to capture the complexity and scope of what is included in this tradition, which includes battles against maldistribution, inequality in process, explicit or systemic discrimination and redress for harm to health and well-being. Kuehn suggests that '[e]nvironmental justice means many things to many people' (2000, p 10681), while Sze and London (2008) hold that 'environmental justice has struggled over the question of definitions' (p 1332) and 'can mean almost anything' (p 1347). A definition that is commonly subscribed to is provided by Bryant who defines environmental justice in the context of 'cultural norms and values, rules, regulations, behaviours, policies and decisions to support sustainable communities, where people can interact with confidence that their environment is safe, nurturing, and protective' (Bryant, 1995, p 589). In simple terms, environmental justice could be expressed as justice underlying decision-making on land-use outcomes.

While many scholars equate the tradition of environmental justice as the conceptual embodiment of the social justice movement in the United States emerging at the end of the 1980s, Bullard, dubbed the 'father of environmental justice', conceptualised these as separate strands with similar end results. In this regard, Bullard's approach to identifying the environmental protection campaign and the environmental justice movement as distinct methods reveals what different designations the subject of justice bring to the discussion. At times, these different approaches have been in competition with one another for resources and outcomes, albeit using different tactics to achieve their aims. But frequently they have learned to collaborate, support and even achieve mutual goals (Bullard, 1993). Scholars commonly suggest that environmental justice is the result of the merging of the conservation movement with the social justice movement with the 'linking of environmental goals to subgroup identity and social justice' (Dawson, 2000, p 22) and often, this is related to the home and to one's community (Hamilton, 1990).

The late 19th-century conservation movement in the United States focused on the preservation of natural resources and the protection of nature in terms of protection of the interests of future generations – not least for resource security, but also for biodiversity and ecosystem integrity. However, aside from an intensive setting aside of land and resources in national parks and national reserves, the environmental movement really had its springboard moment in Rachel Carson's *Silent Spring*, which highlighted the potential future ecosystem collapse that would result from pesticide contamination. While justice for nature is frequently absent in mainstream scholarship of environmental justice it does appear in Victorian ecocriticism studies described as 'early environmental justice' (Hall, 2017, p 7), ecological theory scholars that promote 'putting more environment in environmental justice' (Clark et al, 2007, p 66) and, importantly, Indigenous scholars who suggest that environmental justice includes 'responsibilities toward the environment' (Robyn, 2002, p 213).

While both streams have the intention to improve the conditions of the environment at the heart of their objectives, the motivation stems from entirely different rationalities. In the environmental protection movement, nature is the subject of justice, while in the environmental justice movement, the subject of justice is a particular subset of society. The environmental movement has typically focused on attachment, enjoyment and preservation of land and nature while the environmental justice movement has focused on experiences and results of prejudice in the lived environment and 'framing of environmental issues in terms of discrimination against a particular population' (Dawson, 2000, p 23). Both streams have at their heart the ambition to change the trajectories and outcomes underlying the political economy of environmental decision-making.

Main ideas within the tradition

A key feature of environmental justice is that it begins in the environment and protection of the environment but ends in society with resulting benefits to both nature and humanity. Environmental justice is not limited to domestic contexts only, but expands into the international sphere, with connections between the displacement of environmental harms over international boundaries and the creation of geographic differences (Harvey, 1996) and intertwines 'three limbs of objection – human health, environmental protection and economic security' (Jessup, 2017, p 53). It reveals its key ideas through practices that champion the underdog in asymmetric power relations and support those with disadvantages in economic circumstances (Jessup, 2017). Its main tenets emerge in forms of justice related to distribution, procedure and correction, also including social justice as an underlying objective in its outcomes.

As Bullard notes, a notable difference between these two constituents of environmental justice is in the methods used to achieve their aims. While the environmental protection movement has relied on procedural interruption through legal interventions and lobbying, for example, the environmental justice movement has used direct action methods learned from the civil rights movement, such as public protests, demonstrations and petitions (Bullard, 1993). However, there is increasing cooperation between these two groups with 'technical advice, expert testimony, direct financial assistance, fundraising, research, and legal assistance' (Bullard, 1993, p 26), which resulted in a blurring between the methods and increasing uptake of grassroots movements to use legal interventions. Increasingly there is less division between the methods underlying the two streams and rather more differences between sites of injustice, where, in the United States, activism and the courtroom are common arenas for facilitating change, while within the United Kingdom, for example, it is still based in a growing discourse for non-governmental organisations and government (Agyeman, 2002).

While environmental justice has its roots in local appeals to remedy maldistribution and procedural injustice, there is a growing trend to link the methods, aims and ambitions used in local environmental justice actions to international and global efforts to fight transboundary issues that invoke both environment and social justice along both horizontal and vertical dimensions. While the main taproot of environmental justice is anchored in human-related concerns, in more recent years, and particularly from perspectives within critical environmental justice studies, the notion of non-human and more-than-human aspects has broadened the reach of justice considerations to include non-human entities such as air, water, worms or mountains (Pellow, 2018).

In its horizontal dimensions, environmental justice can be seen as the same types of justice aims in different geographic contexts. Examples of applications of environmental justice in other national contexts can be found in South Africa, where activist initiatives focus on correcting historical dispossession of ancestral homelands once seized during apartheid in the name of nature preservation (McDonald, 2002); in Sweden, where there is a focus on procedural justice in the disruption of the mining permitting processes in Indigenous reindeer herding communities; while in Australia, citizens fought against distributional injustices in the proximity and exposure to toxic waste dumps and chemical fires (Lloyd-Smith and Bell, 2003). In horizontal applications of environmental justice, the forms of justice can be actioned using similar methods and patterns.

In its vertical dimensions, environmental justice becomes global and transnational in its harms and in its corrective ambitions. In this regard, other substantive approaches to justice, such as climate justice or energy justice, can be perceived as corollaries to environmental justice, particularly at the global scale. In global environmental justice, 'environmentalism of the poor' becomes the 'environmentalism of the dispossessed' (Temper, 2014, in Martinez-Alier et al, 2016, p 732). Notable in global environmental justice is that while environmental harms can be local or transboundary (that is from resource extraction by multinational corporations or atmospheric pollution), that the scales of power asymmetries can see an additional degree of removal with decision-making on harms embedded across internationally situated regulatory jurisdictions.

Aspects of environmental justice from a distributive justice perspective are defined as the 'right to equal treatment, that is the same distributions of goods and opportunities as anyone else has or is given' (Dworkin, 1977, p 273, in Kuehn, 2000). In distribution, the achievement of justice is reached through 'fairly distributed outcomes, rather than on the process for arriving at such outcomes' (Kuehn, 2000, p 10684). Issues emerge in distributive aspects of environmental justice because perspectives of just distribution can be contingent on time and place, rather than having a universal application, except in the context of regulated measurements for exposure to harmful substances.

Questions of distribution emerge in whether injustice is situated in the density of maldistribution for some populations or places, over even distribution of unwanted environmental hazards both geographically and across all demographics. In this, there is a tension between environmental justice requiring a good quality of the environment in all places or whether environmental justice requires 'relative deprivation', which, as Helfand and Peyton say, 'suggests that people are concerned about their standing in a community relative to their neighbours rather than about their absolute standard of living' (1999, p 70). The span between these two variations of distribution is based on the difference between conceiving of justice as fairness, and justice as including responsibility towards both man and nature.

Distributive components of environmental justice have both negative and positive features. The negative features emerge in proximity to environmental harms and in adverse effects on health and well-being due to risks and exposure to environmental hazards. In its distribution, environmental justice has a 'revived concern about toxicity and its impact on both people and habitat' (Jessup, 2017, p 56) and is considered in both the proximity to and from environmental harms or amenities or risk of exposure or adverse effects from environmental hazards. The areas purposefully allocated for environmental harm have sometimes been described as sacrifice zones, or an 'area that is considered lost due to environmental degradation and sacrificed for a higher (economic, national security, and so on) purpose' (Skorstad, 2023, p 97).

As a method for evaluating unjust distribution, proximity and exposure to environmental harms, common assessment methods have emerged within the field. These include 'unit based, distance based and exposure/risked based analyses' (He et al, 2019, p 2). These methods are useful for analysing maldistribution in quantitative measures against specific sociodemographic indicators. However, there are concerns that consideration of distributive features only, which is found in a large number of studies of environmental justice, 'ignores questions of causation and agency, and obscures underlying social processes' (Foster, 1998, p 778). These methods are useful for assessing quantitively the various aspects of harm but do less for measuring power asymmetries.

Procedural justice includes the right 'to equal concern and respect in the political decision about how these goods and opportunities are to be distributed' (Dworkin, 1977, p 273, in Kuehn, 2000). In its procedural aspects, concerns of environmental justice scholars linked resulting injustice to a lack of social power (Bullard, 1993) to influence or participate in processes that ultimately result in environmental harms and exposure to environmental hazards. This is highly correlated with race, income, jobs, education, assets and health, and is not related to environmental appropriateness for affected sites (Bullard, 1993). However, although dimensions of justice are related to procedural or participatory processes, it has been flagged that 'increasing community participation is no silver bullet' with real justice relying 'on communicative planning processes to neutralize preexisting power inequalities' (Garrison, 2021, p 7). It is recognised that participation in meetings does not always result in real participation in the decision-making and outcomes.

Researchers and activists who focus on achieving participatory advances in pursuit of environmental justice are 'less concerned about statistical significance or the appropriate unit of analysis', rather they are attempting to 'solve a specific social, economic and/or environmental problem ... to achieve local solutions' (Bryant, 1995, p 600). Increasingly, participatory methods include participatory research or citizen science, which empowers citizens through 'having played a greater role in the decision-making process, will also share the responsibility for outcomes' (Bryant, 1995, p 609). This aspect of environmental justice links research with activism where perspectives of just outcomes are rooted in 'the fairness of procedures leading to the outcome' (Villa et al, 2020, p 330, in Kuehn, 2000). However, even when progress can be made in achieving regulatory controls against polluters, Bryant noted that regulatory controls on environmental pollution still do not fundamentally address 'unequal distribution of power, wealth and income in society' (1995, p 598), which are at the root cause of environmental injustice.

A third feature of environmental justice is found in corrective or retributive forms of justice. Bryant proposes that procedural actions that seek to remedy maldistribution and exposure to the environment is not sufficient and that the remedy lies in 'changes in the structural underpinnings of society that give birth to environmental and social degradation' (Bryant, 1995, p 589). In this, corrective justice emerges as a remedy for environmental injustice and 'involves fairness in the way punishments for lawbreaking are assigned and damages inflicted' and 'attempts to restore the victim to the condition [they] were in before the unjust activity occurred' (Kuehn, 2000, p 10693). While restoration may be impossible in the case of health and well-being, compensation may alleviate suffering or prevent harm to future generations.

As ways to address corrective action in compensation or rectification for environmental harms and hazards, a number of principles have been promoted. One of these principles is the idea that the polluter pays. This has been levied in domestic settlements as a form of financial liability for impact from pollution or toxic spills, and is also considered within global environmental justice, not only for ground or water pollution from isolated events, but also for climate change and atmospheric pollution by greenhouse gases (see Chapter 10). However, it is a challenge that 'a significant amount of climate polluting activity took place over the last few hundred years and was carried out by countless different individuals and businesses' (Coventry and Okereke, 2018, p 364), making specific liability difficult to either determine or enforce.

Within global environmental justice, additional principles of corrective justice suggest that responsibility can be determined based on the ability to pay or that the beneficiary pays. Bullard notes in relation to distribution or correctional aspects of justice that 'the question of who pays and who benefits is central to analysis' (Bullard, 1993, p 21). While he was speaking specifically in regard to environmental racism, which has been at the heart of the environmental justice movement, the distributions of harms and benefits can also be associated with power asymmetries elsewhere, such as with environmental sacrifice zones or green colonialism. In their essence, '[e]nvironmental injustices are instances of not being asked, not being considered, not being recognised and hence, not having an equality of opportunity' (Jessup, 2017, p 62).

The principle of ability to pay has been referred to as the equity principle, but rather than facilitating corrective measures for environmental injustice has 'proven a constant source of disagreement and national rivalry' (Coventry and Okereke, 2018, p 369) in that this principle 'exposes developed countries to financial obligations what would not be politically acceptable' (Coventry and Okereke, 2018, p 370). This is because in an integrated global economy underpinned by the pursuit of growth, expansion and profit, 'it is difficult to realize environmental justice in the competitive market; environmental injustice is the normal state' (He et al, 2019, p 17). These principles, while helping to generate responsibility for environmental hazards, are also difficult to enforce across international boundaries when the site of harm or the benefit of developments falls across different jurisdictions, a difference which is frequently exacerbated by inequalities in geopolitical power (Parks and Roberts, 2007), such as between the Global North and Global South.

Elements of social justice also feature in the consideration of environmental justice in its concentrated 'efforts to bring about a more just ordering of society – one in which people's needs are fully met' (Rodes, 1996, in Kuehn, 2000, p 10698). In the social justice aspects of environmental justice, the environment is understood to be a critical site for 'creat[ing] the conditions for social justice' (Schlosberg, 2013, p 37).

Without a good environment and healthy nature, social justice, including all its components such as economic justice, health justice, energy justice, racial justice, and so on, cannot be achieved.

One concern of social justice in relation to the environment is concerned with the effects of cumulative impact from exposure to environmental harm and 'looks at risk in combination within the complex context of people's lived realities' (Sze and London, 2008, p 1338) and in their exposure pathways. The movement to achieve social justice through environmental indicators has been described as 'integrating environmental concerns into a broader agenda that emphasizes social, racial and economic justice' (Alston, 1990, in Kuehn, 2000, p 10699). Taylor has described the social movement for environmental justice as 'socially constructed claims defined through collective processes' (2000, p 509). In this, there is an opportunity for variation across different communities for defining healthy communities and well-being in their own terms.

Main critiques of the tradition

There are a number of criticisms related to both normative perceptions of environmental justice and of the objectives and methods of the environmental justice movement, more broadly. A dominant critique is that environmental justice efforts are too focused on causality. This criticism is especially targeted towards the US-based environmental justice movement in the need to prove intent and causality in courtroom battles. This is situated as a problem within substantive research and the focus on the maldistribution of environmental harms and hazards, where requiring evidence on deliberate causality 'conclusively proving "racial intent"' (Agyeman, 2002, p 32) has proved difficult to achieve. In contexts beyond the United States, the civil rights context is different and less salient, there is more focus on either class or socioeconomic indicators, yet this variation can be seen to dilute the objectives of activists whose ambition is to make living conditions better for everyone downstream or for, in fact, enhancing 'pre-existing cleavages and increase the potential for conflict' (Dawson, 2000, p 22).

A second feature in this focus on causality is the problem that 'correlation is not causality' (Helfand and Peyton, 1999, p 68), which suggests that even if it can be proven that more substantial environmental harms do in fact exist, it does not mean that they were either intentional, or rather the effect of Ricardian rents. In this it is suggested that it is difficult to prove 'discrimination at the time of siting' or 'market dynamics that lead these groups to locate in areas that are already home to a site' (Helfand and Peyton, 1999, p 69), resulting in weak legal foundations for achieving redress (Kevin, 1997). While both are issues of justice, the latter may be an issue of structural injustice (Young, 2011) rather than specifically of environmental injustice, or as is sometimes inferred, environmental racism. However, this criticism is itself subject to scrutiny and criticism because it implied that exposure to environmental harm is a matter of choice and that the poor are responsible for their own exposure to environmental hazards, rather than victims of injustice through powerlessness.

A third criticism is that environmental justice is stuck in distributive justice. Schlosberg suggests that the scholarship on environmental justice frequently fails to take into account ongoing developments in environmental justice, especially from scholars such as 'Iris Marion Young, Nancy Fraser and Axel Honneth [who] argue that while justice must be concerned with classical issues of distribution, it must also address the processes that construct maldistribution' (2007, p 2). It is also suggested by Foster that empirical research also requires 'analysis of agency and causation in institutional and social processes that lead to distributive outcomes' (1998, p 790). This is relevant in bridging hierarchies of power that impose decision making resulting in maldistribution, not just at local levels, but also across global scales and in bringing capabilities and recognition to questions of environmental justice for both individuals and groups alike.

Another critique is that environmental justice is not either theoretical or methodological enough and is specifically criticised for 'being insufficiently theoretical about racism and how racism actually operates' (Sze and London, 2008, p 1341). This emerges in particular to the relationships between structure, power and geography that condition spatial relationships. This is especially directed towards what is frequently labelled as first-generation research on environmental justice emerging in the 1980s, which although undoubtedly important frontier research, was 'insufficient and inadequate to the tasks of both revealing inequalities and understanding the processes through which these are (re)produced' (Walker, 2009, p 516). In this regard, historical research is an important component in framing the long-term development of inequalities that emerge as the result of various vectors, including political contexts and time periods (Sze and London, 2008). The result is that this has a particular effect of limiting 'social justice claims and ultimately reproduc[ing] a racist social order' (Sze and London, 2008, p 1341). In addition, Bryant criticises the methodology emerging from scientific research on environmental harms where 'politicians, policy makers, or corporate managers decide upon end values', promoting participatory research, which today might be called co-production, as a more 'democratic research process'. In this process Bryant advocates for a spiral production of knowledge where truth is derived through a repetitive cycle of 'planning, acting on the plan, and then observing and reflecting on the results' (1995, p 600).

Environmental justice is also criticised for being too exclusionary. Although interests in the environment and justice have intersected for over a century, environmental justice owes its advent as a field of research or social movement due to the connection between the environment and the civil rights movement in the 1980s. However, when the concept of environmental justice is restricted to a narrow definition of environmental racism, it eliminates vast swathes of affected populations from the pursuit of justice from exposure to environmental harm by 'limiting the types of communities that could make environmental justice claims' (Schlosberg, 2007, p 5). In this regard, Jessup argues that 'if that narrative of environmental justice is adopted, then the community of environmental justice becomes narrow: excluding communities whose justice concern is not distribution or whose vulnerability is not grounded in race, ethnicity or class' (2017, p 49). Bullard also criticises the two different branches of environmental justice for not adequately meeting on issues of 'economic development, social justice and environmental protection' (Bullard, 1993, p 22) and that the environmental protection arm 'has not sufficiently addressed the fact that social inequality and imbalances of social power are at the heart of environmental degradation, resource depletion, pollution and even overpopulation' (Bullard, 1993, p 23), many concerns that are central to their end goals.

A final criticism with growing significance includes that there isn't enough environment in environmental justice where there is frequently a gap between the social and the ecological. It is becoming increasingly apparent that this gap results in an irreconcilable difference as environmental justice expands into the global agenda. In this regard bringing 'the environment back into environmental justice scholarship ... [is] a promising new direction' (Sze and London, 2008, p 1345). It is this gap that critical environmental justice studies address, at least in part, by allocating for the agency of non-human subjects of justice.

Conclusion

There are a number of considerations for the future of environmental justice. First, is that analysis of environmental justice researchers and activists must avoid narrow analyses of environmental harm. At its core, environmental justice is 'a matter of disproportion impact' (Garrison, 2021, p 8) of the distribution of environmental harms and amenities, inequality in decision-making, inadequate recognition of the role of unequal power structures and biases, and lack of measures to compensate or allocate responsibility for harm. It also must encompass 'the fair treatment of people with respect to the execution and application of environmental politics' (Pulido, 2017, p 46). The dynamics of distribution, procedure, correction and recognition are critical in tangent for any evaluations or solutions for environmental injustices. As stand-alone components, they are individually insufficient for developing holistic pathways towards a meaningful environmental justice as the absence of any feature is likely to result in some other form of injustice. This is relevant for both environmental justice at the site of local and specific communities as well as across global scales.

Second, is that if environmental justice is to make substantial contributions to the quality of human life, it must also take the non-human aspects of environmental care and protection as seriously as it promotes social equity. With a focus solely on society and social justice, the instrumental relationship of man to nature will continuously result in the environmental justice movement treating the symptoms, rather than the causes, of environmental injustice. With a focus on social justice only, the fight will remain at the calculations for unjust distribution of proximity and density to environmental hazards or distance to green spaces. It will remain focused on procedural aspects that improve voice and equalise power in the decision-making on these (mal)distributions, meanwhile lacking recognition for the critical and unequal relationship between man and nature or the incapacity of polluters to ever truly compensate for environmental degradation, deterioration of individual health and negative impacts on community well-being. In this regard, environmental justice needs to move beyond its role as a method of social justice and become a method for whole-system justice through the integration of both man and nature and the inclusion of non-human subjects of justice.

Finally, environmental justice must be relevant and applicable across geographic scales and temporalities and 'a global understanding of environmental justice must focus on a "broader set of questions"' (Sze and London, 2008, p 1343), and 'expand in scope to include global processes' (Nelson and Grubesic, 2018, p 8). It is important to reckon with the reality that as problems of refuse, pollution and toxic spills grow in scale and magnitude, transgressing geographic boundaries (that is, ocean plastic pollution and acidification, global warming through increasing concentrations of greenhouse gases, rising sea level and insecurity of water supply), we all, in fact, live downstream to any environmental hazards. While 'some live more downstream than others' (Tarter, 2002), eventually all of the environmental harms and lack of environmental responsibility will come full circle. Proximity will increasingly be a less clear divider for at-risk populations and access to procedural justice may become a moot remedy. In this regard, theoretical advances and empirical application of environmental justice should become transferrable across scale, time and place to remain effective as a method and practice for pursuing justice.

References

Agyeman, J. (2002) Constructing environmental (in)justice: Transatlantic tales. *Environmental Politics*, 11(3), 31–53.

Bryant, B. (1995) Pollution prevention and participatory research as a methodology for environmental justice. *Virginia Environmental Law Journal*, 14(4), 589–613.

Bullard, R.D. (1993) Anatomy of environmental racism and the environmental justice movement. In R.D. Bullard (ed), *Confronting Environmental Racism: Voices from the Grassroots*. Boston, MA: South End Press, pp 15–39.

Clark, W.C., Kates, R.W., Richards, J.F., Mathews, J.T., Meyer, W.B., Turner II et al (2007) Relationships of environmental justice to ecological theory. *Bulletin of the Ecological Society of America*, 88(2), 166–170.

Coventry, P. and Okereke, C. (2018) Climate change & environmental justice. In R. Holifield (ed), *The Routledge Handbook of Environmental Justice*. London: Routledge, pp 362–373.

Dawson, J.I. (2000) The two faces of environmental justice: Lessons from the eco-nationalist phenomenon. *Environmental Politics*, 9(2), 22–60.

Dworkin, R. (1977) *Taking Rights Seriously*. Cambridge, MA: Harvard University Press.

Foster, S. (1998) Justice from the ground up: Distribute inequities, grassroots resistance, and the transformative politics of the environmental justice movement. *California Law Review*, 86(4), 775–841.

Garrison, J.D. (2021) Environmental justice in theory and practice: Measuring the equity outcomes of Los Angeles and New York's 'million trees' campaigns. *Journal of Planning Education and Research*, 41(1), 6–17.

Hall, D.W. (2017) *Victorian Ecocriticism: The Politics of Place and Early Environmental Justice*. Lanham: Lexington Books.

Hamilton, C. (1990) Women, home & community: The struggle in an urban environment. *Race, Poverty & the Environment*, 1(1), 3–13.

Harvey, D. (1996) *Justice, Nature, and the Geography of Difference*. Cambridge, MA: Blackwell Publishers.

He, Q., Wang, R., Ji, H., Wei, G., Wang, J. and Liu, J. (2019) Theoretical model of environmental justice and environmental inequality in China's four major economic zones. *Sustainability*, 11(21). https://doi.org/10.3390/su11215923

Helfand, G. and Peyton, L.J. (1999) A conceptual model of environmental justice. *Social Science Quarterly*, 80(1), 68–83.

Jessup, B. (2017) Trajectories of environmental justice. *Victoria University Law and Justice Journal*, 7(1). https://doi.org/10.15209/vulj.v7i1.1043

Kevin, D. (1997) Environmental racism and locally undesirable land uses: A critique of environmental justice theories and remedies. *Villanova Environmental Law Journal*, 8, 121–160.

Kuehn, R.R. (2000) A taxonomy of environmental justice. *Environmental Law Reporter*, 30, 10681–10703.

Lloyd-Smith, M.E. and Bell, L. (2003) Toxic disputes and the rise of environmental justice in Australia. *International Journal of Occupational and Environmental Health*, 9(1), 14–23. https://doi.org/10.1179/oeh.2003.9.1.14

Martinez-Alier, J., Temper, L., Del Bene, D. and Scheidel, A. (2016) Is there a global environmental justice movement? *The Journal of Peasant Studies*, 43(3), 731–755.

McDonald, D.A. (ed) (2002) *Environmental Justice in South Africa*. Athens and Cape Town: Ohio University Press and University of Cape Town Press.

Nelson, J. and Grubesic, T. (2018) Environmental justice: A panoptic overview using scientometrics. *Sustainability*, 10(4). https://doi.org/10.3390/su10041022

Parks, B.C. and Timmons Roberts, J. (2010) Climate change, social theory and justice. *Theory, Culture & Society*, 27(2–3), 134–66. https://doi.org/10.1177/0263276409359018.

Pellow, D. (2018) *What is Critical Environmental Justice?* Medford, MA: Policy Press.

Pulido, L. (2017) Conversations in environmental justice: An interview with David Pellow. *Capitalism Nature Socialism*, 28(2), 43–53.

Robyn, L. (2002) Indigenous knowledge and technology: Creating environmental justice in the twenty-first century. *American Indian Quarterly*, 26(2), 198–220.

Schlosberg, D. (2007) *Defining Environmental Justice*. Oxford, UK: Oxford University Press.

Schlosberg, D. (2013) Theorising environmental justice: The expanding sphere of a discourse. *Environmental Politics*, 22(1), 37–55.

Skorstad, B. (2023) Sacrifice zones: A conceptual framework for Arctic justice studies? In C. Wood-Donnelly and J. Ohlsson (eds), *Arctic Justice: Environment, Society, Governance*. Bristol: Bristol University Press, pp 96–108. https://doi.org/10.51952/9781529224832.ch007.

Sze, J. and London, J.K. (2008) Environmental justice at the crossroads. *Sociology Compass*, 2(4), 1331–1354.

Tarter, J. (2002) Some live more downstream than others. In J. Adamson, M.M. Evans and R. Stein (eds), *The Environmental Justice Reader: Politics, Poetics, & Pedagogy*. Tucson, AZ: University of Arizona Press, pp 213–228.

Taylor, D.E. (2000) The rise of the environmental justice paradigm: Injustice framing and the social construction of environmental discourses. *American Behavioral Scientist*, 43(4), 508–580.

Villa, C. et al (2020) *Environmental Justice: Law, Policy, and Regulation*, 3rd edn. Durham, NC: Carolina Academic Press.

Walker, G. (2009) Beyond distribution and proximity: Exploring the multiple spatialities of environmental justice. *Antipode*, 41(4), 614–636.

Young, I.M. (2011) *Responsibility for Justice*. New York: Oxford University Press.

10

Climate Justice

Tracey Skillington

Introduction

Climate justice relates to concerns about the inequitable outcomes of climate change for differing peoples, communities, contexts and generations. It aims to achieve greater equity in the distribution of climate related burdens and responsibilities, as well as greater parity of decision-making and human rights protection. More than a normative ideal and a human rights concern, climate justice is also a movement for institutional change, one that documents the differing social, economic, public health and environmental impacts of climate change globally. As sources of climate harms are rarely context-specific, arising as they do from the cumulative effects of atmospheric pollution, rising global temperatures and changes in precipitation patterns across borders, a viable framework of corrective action must be international in scope. Rather than focus on singular aspects of planetary life, a climate justice perspective considers the interrelationship between multiple climate-related forces of change and how they collectively redefine experiences of this world as a space shared in common (see Skillington, 2017; 2019a). For instance, critical insights on the implications of the Earth's changing status as a *terra mobilis* – a planet constitutively in motion and on which we are potentially only guests due to the growing threat of displacement, resource deprivation, illness, death and extinction. While the science of the Anthropocene and its 'great acceleration' provides much in the way of evidence of the exceptional status of this geological age, it is the social sciences and humanities who have fruitfully drawn attention to the ethical and social implications of these changes for planetary life more generally (Skillington, 2015; Chakraborty, 2017). Indeed, it is this emphasis on the global scale of change that most distinguishes a climate from an environmental justice perspective, the latter of which tends to focus more on disputes arising in relation to the natural properties of specific territories (see Schlosberg, 2007; Morea, 2021).

Another key characteristic of a climate justice perspective is its focus on the 'who' as much as the 'what' or the 'how' of climate change events. In the process, all aspects of climate change come to be inflected with a degree of critical normative consciousness. Malm (2016), reflecting on the humanitarian consequences of climate change begs the question of why more serious consideration is not given to the matter of 'who lit this epochal fire?' Along with his colleague, Alf Hornborg, Malm points to several reasons why 'Anthropocene narratives' apportioning blame for climate wrongdoing to the human species as a whole are problematic (see Malm and Hornborg, 2014). Equally, discussions on the relevance of the distributional, procedural, retributive and recognitional components of justice have been criticised for not taking the wider geopolitical aspects of climate change sufficiently into consideration. For instance, efforts to highlight the ongoing impacts of imperial histories of fossil fuel plunder (Carbon Brief, 2021) or the influence of related structures of economic, social and political inequality on the changing dynamics of a warming world (Clark and York, 2005; Moore, 2017). Similarly, criticisms of a failure to connect institutionally embedded patterns of discrimination and value inequality (McNay, 2008) with current climate change disadvantage in justice reasoning (Brugnach et al, 2014). For these authors and more (see also Beck, 2009), it is essential that justice be measured from a standpoint immanent within historically conditioned social practices and in a manner that also incorporates its transformative potential (or capacity for transcendence). From this perspective, current justice relations and levels of climate destruction cannot be considered 'unintended' since they reflect the actions of specific decision-makers committed to preserving asymmetric relations between those who take risks and reap enormous financial benefits, and those who suffer the dire consequences of these decisions (Skillington, 2012). Equally, they cannot be considered unchangeable since their endurance simultaneously embodies the possibility of immanent transcendence.

Common but differentiated responsibilities for climate change

Unlike an earlier, almost exclusive, focus on scientific assessments of climate change risk, from the early 1990s onwards a notable shift occurred in international discourse on climate change. Attention now shifted towards 'common but differentiated responsibilities' for its worst effects. Gaining more attention were the ethical, social, political and economic implications of climate change and their disproportionate effects on the life circumstances of some (for example, see UN, 1992, Article 3). Greater emphasis was placed on the question of justice and empirical applications of an ethic of responsibility, as well as principles of right, equality, and fairness. Those states who contribute most to climate problems in terms of emissions levels and who have gained considerably by doing so, historically speaking, were now openly seen as morally obliged to pay more towards the costs of protecting the climate system. The receipt of benefits in the form of material gains were thought to generate certain duties of responsibility (that is, the beneficiary pays principle) for major polluters and special rights entitlements for affected communities (Shue, 2010). That is, those for whom cumulative pollution-generating activities have gravely depleted the availability of essential resources and decent life chances. The UN (1992) noted the 'unwarranted economic and social costs' being imposed on poor, developing countries (see UN, 1992, p 1) and the growing disparities emerging between peoples in terms of the ability to adapt to climate change (McMichael et al, 2008). Importantly, these early assessments of the unjust aspects of climate change began to highlight in more detail how those who face the highest risks of climate disaster are not only those least well prepared to withstand its worst effects (Brouwer et al, 2007) but are also those least responsible for creating these problems. Among those formally recognised as 'unfairly disadvantaged' in this instance were the communities of tropical and subtropical regions of Africa, Asia and Latin America (for example, see Stern, 2006), as well as the peoples of small island developing states (SIDS) in the Caribbean, the Pacific, the Atlantic, the Indian Ocean, the Mediterranean and South China Seas (Mycoo, 2018). Already, the intensity of storm surges and the proportion of high magnitude tropical cyclones in these regions has increased significantly (Intergovernmental Panel on Climate Change, 2018). Several low-lying Pacific Islands in the Solomon Islands and Micronesia have already been lost, including Kale and Rapita in the northern Solomon Islands. More still are experiencing severe erosion due to steady sea-level rises (Nunn et al, 2017). For instance, the islands of Hetaheta and Sogomou in the northern Solomon Islands where 62 per cent and 55 per cent of island loss respectively has occurred (Sweet et al, 2014). The current life situation of marginalised communities in all of these regions is one affected by a climate of 'total change' where global warming combines with already existing economic, social and cultural challenges, leading to a further expansion of inequalities.

Principles of fairness demand that those left with less than enough to sustain a fruitful and decent existence as a consequence of global climate changes be provided with some form of compensation (that is, the polluter pays principle [see O'Neill, 1986, p 75]). According to Singer (2011, p 190), this reasoning is equivalent to the belief that 'people should contribute to fixing something in proportion to their responsibility for breaking it'. But what happens in circumstances where there is uncertainty as to how far the ecological predicaments of the peoples of climate vulnerable regions have been created by the historical wrongs of larger climate agents? Establishing liability for harms encountered in this instance might prove difficult. Even so, those communities who are most vulnerable insist on a 'backward looking' perspective when determining responsibility for climate change and call on states to 'take responsibility for their actions and their ends' (Harris, 2010). Developed states, on the other hand, highlight various practical difficulties encountered when trying to attribute historical responsibility for climate harms, especially those harms generated at a time when their full effects were not known. For those sceptical of the practical validity of a backward-looking polluter pays principle, the liberal idea of moral responsibility must be specified in terms of free agents carrying out pollution acts in full knowledge of their consequences. In this instance, a polluting agent/state cannot be held legally responsible for their actions if they were unaware of the effects of those actions at the time (that is, displays an 'excusable' degree of ignorance of the effects of their emissions in the pre-1990 period). However, as Shue (1997) points out, while an excusable ignorance argument may limit liability for pre-1990 pollution emissions, it does not eliminate liability for the costs of cumulative harms generated thereafter (for example, a 'time slice approach' [see Nozick, 1974]). It may seem deeply unfair to some that cumulative historical emissions be ignored and we begin only from 1990 levels when determining levels of responsibility for climate change adversities, especially when emissions accrued over centuries continue to dismantle the capabilities of the climate vulnerable to secure basic needs. Greenhouse gas pollution, whose span of affectedness reaches far beyond its point of initiation, is known to have a forcing effect on future climate conditions (NASA, 2017), undermining the capacity of millions to secure the prerequisites needed to live a safe and healthy life. The ecological fate of past, present and future generations is unavoidably connected. Because the destruction that flows from many centuries of emissions almost certainly cannot be reversed, cumulative climate harms generate *pro-tanto* duties to invest in climate change mitigation and adaptation measures. We may assume technology will offer newer generations better ways of coping with the dire effects of this cumulative destruction but in that, we cannot assume such knowledge expertise will also be considered reasonable compensation for what were once avoidable harms (for example, the decision to continue to exploit and invest in fossil fuel economies). Even so, all actors agree, even at some minimal level, on the importance of a transcendence from within a context of formidable destruction and commit to the ideal of a better future.

Climate justice across generations

Unlike principles of responsibility focused on the present, in an intergenerational setting, such principles typically centre on a different kind of collective action problem, one that starts from the need for a deeper framework of climate justice embracing at least several generations (Skillington, 2019b; 2019c).[1] Sceptics point to various practical challenges in trying to actualise such an approach. In particular, how a lack of 'real interaction' across generations (present and future, as well as present and past) prevents a necessary degree of cooperation from emerging to make a multigenerational model of climate justice truly feasible. Typically, cooperation entails a degree of 'give and take' between parties, with one party giving to the other something and receiving something in return. On the surface, the exchange model would indeed appear to create difficulties for a viable intergenerational model of climate justice to succeed, especially when further future generations (those not yet born) are taken into consideration. Traditionally, contract theories of justice (for example, Rawls) take as their primary focus the issue of distribution – how to distribute surplus goods produced by shared productive endeavours (a 'fruits of our labour' argument). Rawls, for instance, argues that it would be unfair to expect present generations to abstain from high polluting activities that generate wealth for the benefit of future peoples, given that the latter cannot reciprocate to the advantage of the presently living. Rawls (1999) refers to this as 'chronological unfairness' where those who come later profit from the wealth-generating actions of their predecessors 'without paying the same price'. However, Rawls does not consider the reverse argument where chronological unfairness is viewed in terms the ecological adversities created knowingly by present-day polluters for future peoples.

In a scenario where distributive justice is seen as intergenerationally and globally applicable, a contract model of justice may still seem morally justified. Even so, support for an intergenerational contract no doubt will continue to encounter resistance given the scale of intertemporal discounting still in operation (for example, short-term policy reasoning on energy options). Intertemporal discounting occurs when the long-term costs of carbon-intensive development pathways are repeatedly played down as a distant concern affecting those who, in not yet being born, lack an identity and a legitimate claim to justice. Deeply ingrained in such reasoning are two assumptions: first, that the near effects of environmental actions are more important, ethically speaking, than remote ones and second, that the effects we produce as individuals are more significant than those we produce as members of wider collectives (for example, generations or state communities). A traditional liberal understanding of the primacy of individual responsibility and rights poses problems in terms of assigning duties for long-term harms. Dead persons cannot be held accountable for historical emissions. States, on the other hand, as collective entities that endure (usually) over time, can be held responsible (Neumayer, 2000). As climate change agents, states do not leave the societal stage in the same way as people do.

Their resource-depletion choices continue to exert an influence for many decades, even centuries, shaping the lives of many people. States, therefore, serve as ideal candidates for initiating a deeper framework of climate justice and a more intergenerationally relevant contract model of just distribution.

Even so, it is still the ecological present that is prioritised in current policy approaches to climate change, leading the global emissions trajectory to continue to soar. Given this scenario, is it still possible to claim that the principles upon which the dominant liberal model of justice is built (freedom, equality, respect for rights, and concern that 'enough and as good' is left for those that follow) are applied without prejudice? Equally, is it still possible to legitimately defend the appropriation of what remains of the global carbon sink on the same grounds as has traditionally been the case (ignorance of its effects)? An effective model of climate justice must address issues of profound irresponsibility, non-accountability and expanding inequalities between generations, regions and, indeed, species (Skillington, 2016; 2019b). In relation to the latter, legal formulations of the rights of nature have received more specific and enforceable consideration in recent years in various court settings and quasi-judicial processes, as a liberal rights discourse continues to evolve.[2] However, more needs to be done in terms of the recognition, protection and assertion of the rights of various planetary species and bio-geophysical systems as essential correlatives to ensuring the liveability, integrity and regeneration of the natural world for present and future generations.

Addressing climate wrongdoing from within a deeper justice framework

For those advocating for new standards of climate justice across the international stage, a key concern is how change can be actualised in a more institutionally consequential manner? Apart from legal changes, theorists such as Pogge (2008) and Nussbaum (2006) have supported the implementation of a series of economic instruments to move societies away from carbon-intensive models of development. One proposal is a global resource tax system designed to target escalating rates of CO_2 emissions, especially from fossil fuel consumption. Pogge (2008, pp 202–221) suggests that taxation funds accrued on natural resource consumption be redistributed to less affluent states through a Global Resource Dividend as a form of compensation for environmental damages and to fund new climate adaptation strategies. The primary aim of this system of taxation is to ensure that those who continue to over-appropriate scarce resources and prosper greatly by doing so, pay for harms they impose on others. Other theorists, however, have questioned whether a pollution tax can effectively reduce overall emissions levels (for example, Caney, 2014). In a similar vein, Brooks (2019) queries whether it is really possible for states to tax their way to greater climate justice and secure a more sustainable future using a 'business as usual' approach. Taxes may be a standard feature of modern state economics but are not necessarily an effective deterrent to excessive fossil fuel consumption (Collins, 2019). Ultimately, it is not a pollution tax that will discourage the rich from over-spending carbon budgets and causing global greenhouse gas emissions to soar. As there is no overdraft privilege available in this instance (Caney and Hepburn, 2011), a broader range of regulatory instruments is needed (green incentive schemes) to keep global emissions within safe limits. But who will insist a cap on emissions levels will be imposed and a fair taxation regime implemented? An open and inclusive governance structure is vital to ensure unsustainable carbon-intensive development paths are discontinued in the near future and a deeper justice framework adopted instead. The latter is both diachronic (concerned with actions performed over time) and synchronic (concerned with the transformative action potentials of the present) in focus and, in that, embraces a long-term perspective on the health and well-being of global communities. Health in this instance relates not only to the physical but, also, the mental and civic well-being of communities, all of which, studies show, are detrimentally impacted by climate change (for example, *The Lancet* Countdown on health and climate change, 2021).

In *Our Common Agenda* (2021) UN Secretary General António Guterres emphasises the importance of grasping the present as 'the time to renew the social contract between governments and their people' and acknowledge the limitations of existing regulatory frameworks that, to date, have failed to address 'challenges [that] are interconnected across borders and all other divides'. For instance, the rapid melting of historic glaciers, the destruction of ancient forests, the critical loss of plant and animal species, as well as various low-lying territories. Such losses have a strong non-compensatory component to them due to their rich cultural, social and psycho-emotional value and largely preventable status. Only when met with a 'multilateral response' on the part of 'interconnected communities' and a cosmopolitan analytic perspective brought to bear on the significance of such losses to all members of the global commons, can a more effective response be devised to the same. Apart from the UN, other key exponents of this position include Held (2009), Beck (2006) and Caney (2001; 2005), all of whom highlight a need to reflect pragmatically on the importance of securing a lasting peace under conditions of growing natural resource scarcity and consider how members of interconnected communities might encounter one another in situations of increasing resource constraint. At present, such issues are usually met with one of two responses: first, a philosophy of 'each to their own' in a world where there is no global sovereign or supreme arbitrar of conflicts over growing resource shortages. In such a context, there is no equality of power among states and, therefore, it is up to each state to protect the interests of its own members. The preference, therefore, is to prioritise special obligations to state compatriots. Indeed, for theorists such as Miller (2010), acting special obligations to state citizens is the most effective means of bringing about global justice (see also Goodin, 1987). The rights claims of compatriots are granted greater moral weight when deciding how scarce resources will be distributed. A second response, usually provoked in reaction to the first, is that the global reach of climate change undermines the validity of claims that duties to compatriots should always take precedence over those to distant others, or that responsibilities ought to stop firmly at state borders (see also Miller, 2007; Vanderheiden, 2008). As a life-supporting commons, the climate system is inherently cosmopolitan in ways that necessitate a more common earth reasoning. Regardless of their geographic origins, rising greenhouse gas emissions have an impact everywhere, causing particular and universal ecological fates to continuously collide. This fact alone requires that justice be allocated in a more globally relevant manner (Baer, 2010).

Climate justice and human rights

As associates of local, regional, national and international communities affected by climate change, all peoples everywhere have a legitimate claim over decision-making on these issues. Indeed, it is our associative relationship with each other, as fellow earth dwellers, that forms the strongest ethical basis for cooperation. Second, the a priori condition of our possession of the earth's territories and their finite resources (that in their original state belong to all [Kant, 1996 (1797)]), is that we share them with others (including distant others). Theorists such as O'Neill (2001) and Gardiner (2011) define climate justice in these terms, as governed by a cosmopolitan order of rights and duties of care, operating within a framework of regional, national and transnational reciprocity (O'Neill, 2001). This may be referred to as the 'all subjected' model of climate justice where obligations to provide assistance to those currently or likely in the future to be without adequate food, water, energy or place security be specified legally as a matter of allocated justice for all as members of the global community. Actualising rights or shared responsibilities to protect the planetary system for all inhabitants, however, requires that the following principles of cosmopolitan justice be applied in a more consequential manner (for a more detailed discussion on these principles, see Chapter 3). First, that the equal moral worth and dignity of all persons, referred to by Nussbaum (2011) as the basis upon which a minimum threshold of human functioning can be applied, be made more institutionally relevant (see Chapter 8 for a further discussion on the Capabilities Approach). Second, that the principle of human dignity guide concrete applications of the right to safe haven, security, freedom of movement, life, self-determination, and so on, and protect opportunities open to all to live sustainably in contexts where significant climate adversities will prevail (growing risks to land, ocean, coastline and freshwater ecosystems and related losses [see IPCC, 2022]).

Although broadly supported in international discourse on climate change mitigation, cosmopolitan principles are seen to align with real empirical content only when they are situationally applied and explored in relation to concrete experiences of climate adversity. Especially since 2007, those communities particularly vulnerable to climate change, for example, SIDS, have sought to test the validity of the international community's commitments to universal rights. SIDS demand that the currency of international justice be specified evaluatively in terms of the diminishing capacities of their communities to withstand the effects of *globally sustained* climate destruction. As these actors point out, opportunities to adapt to the challenges posed by rising sea levels, more severe storm surges and other climate-related events, are actively curtailed by ecological, political and economic forces largely beyond their control. The distributional, procedural and recognitional aims of climate justice, therefore, need to be defined in ways that take account of this more basic injustice (or what Fraser [2010] refers to the facts of 'abnormal justice'). In November 2007, representatives of the Maldives, Tuvalu and a number of other SIDS requested that the Conference of the Parties seek the cooperation of the Human Rights Council, the chief intergovernmental human rights body in the United Nations, and the Office of the High Commissioner for Human Rights, to bring these issues more to international attention. In highlighting how climate change threatens the human rights of climate vulnerable peoples and, further, how their plight is actively shaped by the decisions of major climate players, these actors bring much-needed clarity to bear on the kind of actions needed to ensure normatively agreed minimum thresholds of responsibility and duties of care are respected (Caney, 2010). This is achieved first through an intrinsic justification for a human rights approach that respects all persons' mutual humanity. Second, one that offers an instrumental justification for a human rights approach that enables each individual to enjoy certain fundamental goods, including freedom of autonomy, good health, freedom from hunger and a decent standard of living. To deprive others of the possibility of meeting these basic needs through failed climate mitigation programmes is to treat them without respect and deny them the basic elements of a decent life (food, water and land security, health, prosperity and a safe future).

In the years since the Office of the High Commissioner for Human Rights published its first report on the relationship between climate change and human rights (January 2009) the climate crisis has come to be examined more centrally through a human rights frame (Skillington, 2012: 1205; OHCHR, 2015), in addition to an economic (for example, OECD, 2015) and cultural rights perspective (for example, OHCHR, 2020). However, it is worth noting that the move towards a rights approach to climate justice did not proceed smoothly at first. Instead, it met with considerable political resistance, especially from among more powerful players. For instance, the United States, fearing a stronger official recognition of the linkages between climate change and human rights violations would bolster the case for further unwanted 'extra-territorial' legal regulations, initially rejected this move (see US submission to the UN Office of the High Commissioner for Human Rights, 2009). Climate change, it noted, is 'one of many natural and social phenomena that may affect the enjoyment of human rights' and, therefore, cannot be singled out as 'the cause' of human rights violations, particularly those arising internationally. Restricting resource rights eligibility to 'legitimate' claimants, particularly those with a legal contractual right to precious reserves of minerals, oil, gas, seeds, forests, arable lands, and so on, and striking 'a balance' between environmental harm and the benefits of the activities causing it were asserted instead as primary concerns.

Causing particular angst for the United States were the efforts being made legally to identify a 'collective or self-standing right to a safe and secure environment'. It noted in its submission to the Office of the High Commissioner for Human Rights, for instance, how this right would almost certainly be used as a 'political or legal weapon' against the United States.[3] For instance, the petition submitted a few years earlier by the Inuit, under the auspices of the Inuit Circumpolar Conference (the Inuit Circumpolar Council since 2006), to the Inter-American Commission of Human Rights, claiming that the United States had violated the rights of the Inuit people to life, food and culture by failing to refrain from actions that would decrease US CO_2 emissions. A similar case was taken by the Arctic Athabascan Council against the Canadian state for violating Athabascan rights through its air pollution practices, especially its contribution to high levels of black carbon, widely considered an important driver of Arctic climate change due to its effects on snow/ice albedo. Although both cases were unsuccessful due to a lack of sufficient evidence to prove categorically the traceability of these harms, they did, nonetheless, raise the profile of a human rights approach to climate justice considerably and the importance of holding states and corporate actors accountable for climate-related harms (principle 3, accountability) in domestic and international court settings. Considering that by February 2020, 1,143 legal cases had been initiated against various federal state government agencies in the United States (see de Wit et al, 2020), it would seem that the concerns of the United States were, indeed, well founded, even if its desire to restrict the legal relevance of a human rights discourse on climate change proved not to be fully justified. Instead, citizens have come increasingly to insist that processes of decision-making on climate change issues be made more inclusive (cosmopolitan principle 4, 5 and 6), dedicated to the avoidance of serious harm (principle 7), consistent with a stewardship of resources that are non-substitutable and conducted in ways respectful of pluriversal worldviews. The validity of these demands was acknowledged recently by the Human Rights Council (in October 2021) when, for the first time, it appointed a Special Rapporteur on human rights and climate change and recognised the right of all peoples to a clean, healthy and sustainable environment. Similarly, the ruling of the UN Child Rights Committee (11 October 2021) that states bear cross-border responsibilities for the harmful impacts of climate change on children's rights. All represent positive steps forward in efforts to frame climate change fundamentally as a human rights justice concern.

Conclusion

This chapter considered some of the main arguments raised in the climate justice literature in recent decades regarding issues of equity in the distribution of the burdens of climate change and responsibilities to protect those most vulnerable to its worst effects. Further, it noted the relevance of both contemporary and historical acts of climate harm to questions of recognition, compensation, accountability and democratic reform. As many complex moral ethical debates are raised by these issues, a justice as fairness approach requires that debate begins and ends with the experiences, insights, needs and circumstances of all affected communities. When 'inclusive' 'participatory' procedures do not begin from this vantage point but rather from the viewpoints of those whose actions ultimately serve to debase the constitutive principles of participation, they wrongly serve only as mechanisms to relay information to communities on decisions made elsewhere. Participation in this instance becomes a form of epistemic injustice – participation in name but not practice. Fairer systems of decision-making, deliberation and enforcement are required, where states continue to be key agents in the coordination and funding of climate change measures (mitigation, adaptation, conservation, deliberation, and so on) but additional layers of governance are operationalised at the local level and, simultaneously, at the transnational level to enable and empower communities at all levels in their efforts to address the specific needs of the climate vulnerable (by generating greater knowledge, expertise, funding, community support, and so on).

Assisting 'all those subjected' to what are collectively manufactured climate adversities must be the primary concern of climate justice deliberation moving forward (Dryzek et al, 2019). Proposals here include a more consistent use of citizen assemblies, juries, discourse chambers and deliberative mini publics to resolve contentious issues around resource management or sustainable energy use and ensure more positive climate outcomes for the future. The aim is to strengthen communities' capacities to reflexively shape the viability of new sustainability and democracy-building projects moving forward. All are intended as a means of empowering communities to be positive influencers overs the future path of climate change and ensure climate justice, as a negotiated truth, remains present in deliberation on related environmental, social, political and economic developments (Dryzek et al, 2019). It is only when publics are enabled in their social and civic capabilities to be 'positive climate influencers' in this moment of 'last opportunity' to avoid irreversible climate freefall (IPCCC, 2022), that such aspirations can become a reality. What is critical, however, is that more ambitious deliberative structures and legal applications of universal principles be recognised and enforced in ways that support mutual advantage and a cosmopolitan model of relationality, highlighting interconnections between peoples, regions, climate actions and outcomes, be explored more thoroughly in terms of its practical relevance (Skillington, 2022). Certainly, there is a growing appetite for change. If the recent tragedies of war in Ukraine have taught the international community anything, it is that a spirit of solidarity and the capacity for cooperative action remain strong. As the biggest threat now facing the world, climate change necessitates an equally committed response in the name of justice for all.

Notes

[1]   The basic idea of contract theory is that contract arrangements should support a 'mutually agreeable reciprocity or cooperation between equals' (see Darwall, 2002: 1). For instance, a state constitution.

[2]   For example, the decision of the New Zealand Parliament in 2017 to finalise the Te Awa Tupua Act appointing two guardians of the Whanganui river, one representative of the Maori Indigenous people and the other a representative of the Crown, reconciling two very different worldviews. In the United States, the City Council of Pittsburgh, Pennsylvania unanimously passed an ordinance in 2010 recognising the Rights of Nature as part of an effort to ban shale gas drilling and fracking. In 2019, the city of Toledo, Ohio, adopted the Lake Erie Bill of Rights, a municipal law that grants the lake legal personhood. Similarly, in February 2021, the Innu Council of Ekuanitshit and the Minganie Regional County Municipality recognised the legal rights of Canada's Magpie River.

[3]   United States (2009). For further discussion, see Skillington (2012).

References

Baer, P. (2010) Adaptation to climate change: Who pays whom? In S. Gardiner, S. Caney, D. Jamieson and H. Shue (eds), *Climate Ethics: Essential Readings*. Oxford, UK: Oxford University Press, pp 247–262.

Beck, U. (2006) *The Cosmopolitan Vision*. Cambridge, UK: Polity.

Beck, U. (2009) *World at Risk*. Cambridge, UK: Polity.

Brooks, T. (2019) Climate change ethics and the problem of end-state solutions. In T. Brooks (ed) *The Oxford Handbook of Global Justice*. Oxford: Oxford University Press, pp 241–258.

Brouwer R., Akter, S., Brander, L. and Haque, E. (2007) Socioeconomic vulnerability and adaptation to environmental risk: A case study of climate change and flooding in Bangladesh, *Risk Analysis*, April, 27(2): 313–326. doi: 10.1111/j.1539-6924.2007.00884.x

Brugnach, M., Craps, M. and Dewulf, A. (2014) Including indigenous peoples in climate change mitigation: Addressing issues of scale, knowledge and power. *Climatic Change*, 140(1), 19–32.

Caney, S. (2001) Cosmopolitan justice and equalising opportunities. *Metaphilosophy*, 32(1–2), 113–134.

Caney, S. (2005) Cosmopolitan justice, responsibility and global climate change. *Leiden Journal of International Law*, 18(4), 747–775.

Caney, S. (2010) Climate change, human rights, and moral thresholds. In S. Gardiner, S. Caney, D. Jamieson and H. Shue (eds), *Climate Ethics: Essential Readings*. Oxford, UK: Oxford University Press, pp 163–177.

Caney, S. (2014) Two kinds of climate justice: Avoiding harm and sharing burdens. *The Journal of Political Philosophy*, 22(2), 125–149.

Caney, S. and Hepburn, C. (2011) Carbon trading: Unethical, unjust and ineffective. GRI Working Papers 49, Centre for Climate Change Economics and Policy & the Grantham Research Institute on Climate Change and the Environment. https://www.lse.ac.uk/granthaminstitute/wp-content/uploads/2011/06/WP49_carbon-trading-caney-hepburn.pdf

Carbon Brief (2021) Which countries are historically responsible for climate change? 1850–2021. https://www.carbonbrief.org/analysis-which-countries-are-historically-responsible-for-climate-change

Chakraborty, D. (2017) The future of the human sciences in the age of humans: A note. *European Journal of Social Theory*, 20(1), 39–43.

Clark, B. & York, R. (2005) Carbon metabolism: Global capitalism, climate change, and the biospheric rift. *Theory and Society*, 34(4), 391–428.

Collins, L. (2019) Riots and trade wars: Why carbon taxes will not solve the climate crisis. *Recharge*, 31 October. https://www.rechargenews.com/transition/riots-and-trade-wars-why-carbon-taxes-will-not-solve-climate-crisis/2-1-694555

Darwall, S. (ed) (2002) *Contractarianism/Contractualism*. Oxford, UK: Wiley Blackwell.

de Wit, E., Seneviratne, S. and Calford, H. (2020) Climate change litigation update. Norton Rose Fulbright. https://www.nortonrosefulbright.com/en/knowledge/publications/7d58ae66/climate-change-litigation-update#section1

Dryzek, J.S., Bächtiger, A., Chambers, S., Cohen, J., Druckman, J.N., Felicetti, A. et al (2019) The crisis of democracy and the science of deliberation. *Science*, 363(6432), 1144–1146.

Fraser, N. (2010) *Scales of Justice: Reimagining Political Space in a Globalising World*. New York: Columbia University Press.

Gardiner, S. (2011) *A Perfect Moral Storm: The Ethical Tragedy of Climate Change*. Oxford, UK: Oxford University Press.

Goodin, R.E. (1987) What is so special about our fellow countrymen? *Ethics*, 98(4), 663–686.

Harris, P. (2010) *World Ethics and Climate Change: From International to Global Justice*. Edinburgh: Edinburgh University Press.

Held, D. (2009) Restructuring global governance: Cosmopolitan democracy and the global order. *Millennium Journal of International Studies*, 37(3), 533–547.

Intergovernmental Panel on Climate Change (IPCC) (2018) Global warming of 1.5°C. An IPCC Special Report on the impacts of global warming of 1.5°C above pre-industrial levels and related global greenhouse gas emission pathways, in the context of strengthening the global response to the threat of climate change, sustainable development, and efforts to eradicate poverty, V. Masson-Delmotte, P. Zhai, H.-O. Pörtner, D. Roberts and J. Skea, et al (eds). https://www.ipcc.ch/site/assets/uploads/sites/2/2019/06/SR15_Full_Report_High_Res.pdf

IPCC (2022) WGII sixth assessment report. Geneva, Switzerland: IPCC. https://www.ipcc.ch/report/ar6/wg2/

Kant, I. (1996 [1797]) The doctrine of right. Part I of the *Metaphysics of Morals in Practical Philosophy*, translated and edited by M. Gregor. Cambridge, UK: Cambridge University Press.

Lancet, The (2021) Countdown on health and climate change. *The Lancet*, 398(10311), 1619–1662. https://www.thelancet.com/journals/lancet/article/PIIS0140-6736(21)01787-6/fulltext

Malm, A. (2016) Who lit this fire? Approaching the history of the fossil fuel economy. *Critical Historical Studies*, Fall, 215–248.

Malm, A. and Hornborg, A. (2014) The geology of mankind? A critique of the Anthropocene narrative. *Anthropocene Review*, 1(1), 62–69.

McMichael, A.J., Friel, S., Nyong, A. and Corvalan, C. (2008) Global environmental change and health: impacts, inequalities, and the health sector. *BMJ*, 336: 191–194.

McNay, L. (2008) The trouble with recognition: Subjective, suffering and agency. *Sociological Theory*, 26(3), 271–296.

Miller, D. (2007) *National Responsibilities and Global Justice*. Oxford, UK: Oxford University Press.

Miller, D. (2010) Cosmopolitanism. In G. Wallace Brown and D. Held (eds), *The Cosmopolitan Reader*. Cambridge, UK: Polity, pp 377–392.

Moore, J.W. (2017) The capitalocene, part 1: On the nature and origins of our ecological crisis. *Journal of Peasant Studies*, 44(33), 594–630.

Morea, J.P. (2021) Environmental justice, well-being and sustainable tourism in protected area management. *Journal of Ecotourism*, 20(2), 1–20.

Mycoo, M.A. (2018) Beyond 1.5°C: vulnerabilities and adaptation strategies for Caribbean Small Island Developing States. *Regional Environmental Change*, 18(12), 1–13.

Neumayer, E. (2000) In defence of historical accountability for greenhouse gas emissions. *Ecological Economics*, 33(2), 185–192.

Nozick, R. (1974) *Anarchy, State and Utopia*. New York: Basic Books.

Nunn, P. D., Kohler, A. and Kumar, R. (2017) Identifying and assessing evidence for recent shoreline change attributable to uncommonly rapid sea-level rise in Pohnpei, Federated States of Micronesia, Northwest Pacific Ocean. *Journal of Coastal Conservation*, 21(6), 719–730.

Nussbaum, M. (2006) *Frontiers of Justice: Disability, Nationality, Species Membership*. Cambridge, MA: Harvard University Press.

Nussbaum, M. (2011) The central capabilities. In *Creating Capabilities: The Human Development Approach*. Cambridge, MA: Harvard University Press, pp 17–45.

Office of the High Commissioner for Human Rights (OHCHR) (2009) Observations by the United States of America on the relationship between climate change and human rights. https://www.ohchr.org/sites/default/files/Documents/Issues/ClimateChange/Submissions/USA.pdf

OHCHR (2015) Understanding human rights and climate change. Submission to the 21st Conference of the Parties to the United Nations Framework Convention on Climate Change, Paris.

OHCHR (2020) Report on climate change, culture and cultural rights. A/75/298. https://www.ohchr.org/en/documents/thematic-reports/a75298-report-climate-change-culture-and-cultural-rights

O'Neill, O. (1986) *Faces of Hunger. An Essay on Povert, Development and Justice*. London: Allen & Unwin.

O'Neill, O. (2001) Agents of Justice. *Metaphilosophy*, 32(1-2), 180–195.

Organisation for Economic Co-operation and Development (OECD) (2015) Adapting to the impacts of climate change: Policy perspective. Paris: OECD. https://search.oecd.org/environment/cc/Adapting-to-the-impacts-of-climate-change-2015-Policy-Perspectives-27.10.15%20WEB.pdf

Pogge, T.W. (2008) *World Poverty and Human Rights*, 2nd edn. Cambridge, UK: Polity.

Rawls, J. (1999 [1971]) *A Theory of Justice*, revised edn. Cambridge, MA: Belknap Press.

Schlosberg, D. (2007) Reconceiving environmental justice: Global movements and political theories. *Environmental Politics*, 13(3), 517–540.

Shue, H. (1997) Subsistence emissions and luxury emissions, In R.L. Revesz (ed.) *Foundations of Environmental Law and Policy*. Oxford: Oxford University Press, pp 322–329.

Shue, H. (2010) Global environment and international inequality. In S. Gardiner, S. Caney. D. Jamieson and H. Shue (eds), *Climate Ethics: Essential Readings*. Oxford: Oxford University Press, pp 101–111.

Singer, P. (2011) One atmosphere. In S. Gardiner, S. Caney, D. Jamieson and H. Shue (eds), *Climate Ethics: Essential Readings*. Oxford: Oxford University Press, pp 181–199.

Skillington, T. (2012) Climate change and the human rights challenge: Extending justice beyond the borders of the nation state. *The International Journal of Human Rights*, 16(8), 1196–1212.

Skillington, T. (2015) Climate justice without freedom: Assessing legal and political responses to climate change and forced migration. *European Journal of Social Theory*, 18(3), 288–307.

Skillington, T. (2016) Reconfiguring the contours of statehood and the rights of peoples of disappearing states in the age of global climate change. *Social Sciences*, 5(46), 1–13.

Skillington, T. (2017) *Climate Justice & Human Rights*. New York: Palgrave.

Skillington, T. (2019a) *Climate Change and Intergenerational Justice*. London: Routledge.

Skillington, T. (2019b) A deeper framework of cosmopolitan justice: Addressing inequalities in the era of the Anthropocene. In G. Delanty (ed), *International Handbook of Cosmopolitanism Studies*, 2nd edn. London: Routledge, pp 383–394.

Skillington, T. (2019c) Changing perspectives on natural resource heritage, human rights and intergenerational justice. *The International Journal of Human Rights*, 23(4), 615–637.

Skillington, T. (2022) A relational view of responsibility for climate change effects on the territories and communities of the Arctic. In C. Donnelly-Wood and J. Ohlsson (eds), *Arctic Justice*. Bristol, UK: Bristol University Press, pp 36–50.

Stern, N. (2006) *The Economics of Climate Change: The Stern Review*. Cambridge: Cambridge University Press.

Sweet, W., Park, J., Marra, J., Zervas, C. and Gill, S. (2014) Sea level rise and nuisance food frequency changes around the United States. NOAA Technical Report NOS CO-OPS 073, National Oceanic and Atmospheric Administration, Silver Spring.

United Nations (1992) United Nations framework convention on climate change. New York: United Nations. https://unfccc.int/resource/docs/convkp/conveng.pdf

United Nations Office of the High Commissioner for Human Rights (2009) Report on human rights and climate change. https://www.ohchr.org/Documents/Issues/Environment/A.HRC.31.52_AEV.docx

United Nations Secretary General (2021) *Our Common Agenda: Report of the Secretary-General*. New York. https://www.un.org/en/content/common-agenda-report/assets/pdf/Common_Agenda_Report_English.pdf

Vanderheiden, S. (2008) *Atmospheric Justice: A Political Theory of Climate Change*. Oxford, UK: Oxford University Press.

11

Energy Justice

Roman Sidortsov and Darren McCauley

Introduction

The concept of energy justice transpired out of recognition by practitioners and scholars of the inequalities, inequities, insecurities, and other moral wrongs and ethical wrongs created in the course of production, transportation and use of energy (Sovacool et al, 2014; Sidortsov et al, 2019). This is a young but rapidly emerging field with its first attempts to develop comprehensive frameworks, conceptions and principles dating back to the early 2010s. At the centre of energy justice is energy as a particular type of good that is instrumental and a prerequisite to human flourishing in the contemporary world.[1] There is not a single country that can claim to be capable of maintaining let alone developing its economy and society without sufficient supply of modern energy and there are not many communities whose members can thrive without access to energy services.

Energy is supposed to be a means to human flourishing through the provision of energy services. However, in reality, it tends to dominate and determine the ends while creating vast and numerous inequities in the process. Oil wars, community displacement, pollution and the looming climate catastrophe are just a few examples of mounting sources of insecurity, inequities and injustice. Correspondingly, the primary goal of energy justice is addressing both the causes and effects of such insecurities, inequities and injustices across multilevel energy systems through exposing and analysing them, as well as developing pathways and solutions for more just production, transportation and use of energy. This does not mean that energy justice scholars are not interested in developing an energy justice theory and contributing to the development of fundamental schools and applied theories of justice. In fact, even a brief survey of scholarly literature, some of which are noted in this chapter, shows several proposed theoretical frameworks and significant contributions to understanding distributive, cosmopolitan, procedural and recognition aspects of justice.

Historical development

The term 'energy justice' first appeared in academic literature in the 2000s in Guruswamy's (2010) article 'Energy justice and sustainable development'. Guruswamy largely equates energy justice with the concept of energy poverty but does not further develop and ultimately define it. Hall (2013) focuses on the possibility of what he calls an ethical consumption, including consumption of energy, while remarking on the lack of an accepted definition of energy justice. The particularly multifaceted nature of energy studies leads to definitional contestation. The next wave of energy justice scholarship explored the foundations of the concept. Heffron and McCauley (2014), Sovacool and Dworkin (2014), Guruswamy (2016), Jenkins et al (2016), and many others, ground the concept in the application of the different aspects and forms of justice that they deem instrumental to an energy system's operations. Jones et al (2015) take a less direct approach. They go through four assumptions to build their conceptual foundation via reconciliation of insights from applicable justice schools with the aforementioned unique characteristics of energy as a good. They develop two principles, affirmative and prohibitive (discussed in more detail in what follows), which they use as the foundation and starting point of analysis for any energy justice problem. Sidortsov et al (2019) term these two approaches respectively 'system' and 'foundational'. This designation is not indicative of analytical superiority of one over another or better suitability for providing solutions to energy centric problems. The chief purpose of the designation is to underscore two divergent pathways that scholars take to deciphering energy justice. To date, the system approach received much wider recognition than the foundational approach.

Energy justice has been shaping out to be a truly energy-first and discipline-second area of scholarship. It did not emerge from a single discipline, university or department. The knowledge of and focus on the energy sector and systems have united energy justice scholars.[2] Often, energy justice is built upon existing research agendas across social sciences in the form of environmental and climate justice (Walker and Bulkeley, 2006; Bickerstaff and Agyeman, 2009; Walker, 2009; Barrett, 2012; Bulkeley et al, 2013), as well as more recent developments in energy poverty (Bouzarovski and Herrero, 2016; García-Ochoa and Graizbord, 2016) and energy vulnerability research (Middlemiss and Gillard, 2015; Bouzarovski et al, 2017a; 2017b). Energy justice engages strongly with geography, legal studies (Guruswamy, 2015), business (Hiteva and Sovacool, 2017), political science (Jenkins et al, 2016), engineering (Heffron and McCauley, 2014) and other disciplines. Because the first attempts to conceptualise energy justice occurred less than a decade ago, analysing its historic development at length is premature. However, with the ever-increasing volume and breadth of energy justice scholarship, the empirical foundation for such analysis is not far away.

The need for historic analysis, reflection and, perhaps, rethinking of the role of energy justice in decision-making is sharpened by several ongoing and emergent crises: climate, energy and supply chain. This differs to just transitions research where it has already played a key role in shaping international policy making as outlined in Chapter 15. Energy justice scholars have established that there is an ethical and moral deficiency of the status quo and outlined many instances, processes and places where injustice occurs. For instance, Sovacool et al (2014) centre their work on instances of injustice that they group into temporal, economic, sociopolitical, geographic and technological dimensions. In general, there is little disagreement among energy justice scholars about the prevalence, impact and significance of injustice in the energy sector. However, attempts to conceptualise different approaches to analysing, pre-empting, mitigating and remedying energy injustices have only begun. For this reason, we focus on the aforementioned system and foundational approaches to deciphering energy justice.

Approaches to defining energy justice

The *system* approach to energy justice builds directly on mainstream theories of justice that the proponents of the approach deemed to be integral to energy sector operations. This is a major point of departure from just transition (Chapter 15), which does not base itself on systems thinking. Unlike energy justice, it places its core focus on transitioning away from carbon-intensive fuels. Thus, the approach aims to address an energy justice problem largely with an already existing arsenal of conceptual and theoretical tools. McCauley et al (2013) premise energy justice on three central tenets: distribution, procedural and recognition justice. Sovacool and Dworkin (2014) and Sovacool et al (2016) tap into a larger pool of justice forms and concepts and develop eight core principles through which they define energy justice. These include availability, affordability, due process, transparency and accountability, sustainability, intragenerational equity, intergenerational equity, and responsibility (Sovacool et al, 2016).

These two conceptual springboards have led to further advancement of the system approach. Heffron and McCauley (2014) explore how the concept applies across the energy life cycle and system (see Figure 1). Heffron and McCauley (2017) also add cosmopolitan justice to their framework as depicted in Figure 1 because of the global nature of the production, conversion, delivery and use of energy. To remedy past injustices, they bring restorative justice into their framework as well. Heffron and McCauley place

Figure 1: Energy justice conceptual framework developed by Heffron and McCauley (2017)

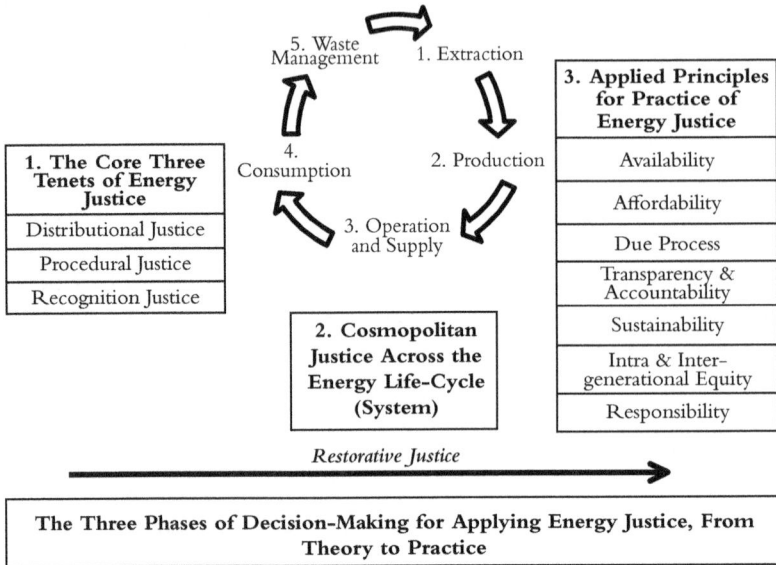

		3. Applied Principles for Practice of Energy Justice
		Availability
1. The Core Three Tenets of Energy Justice	2. Production	Affordability
Distributional Justice		Due Process
Procedural Justice		Transparency & Accountability
Recognition Justice		Sustainability
		Intra & Inter-generational Equity
	2. Cosmopolitan Justice Across the Energy Life-Cycle (System)	Responsibility

5. Waste Management   1. Extraction

4. Consumption

3. Operation and Supply

Restorative Justice

The Three Phases of Decision-Making for Applying Energy Justice, From Theory to Practice

Source: Heffron and McCauley (2017) © 2017 Elsevier Ltd. Reproduced with permission of the Licensor through PLSclear.

the energy life cycle at the core of their framework to ensure that there is an increased understanding of shared obligations by all actors in the energy sector regardless of where individual decisions are made.

Analysis under this framework begins with examining an energy inequity problem (for example, community displacement due to the construction of a hydroelectric dam) vis-à-vis the core tenets (distribution, procedural and recognition justice) to determine its ontology or ontologies. The next step is to broaden the scope to place the problem within the energy life cycle and/or energy system and its global interdependencies and issues. The final step is employing the applied principles for guidance on practical action and restorative justice as a condition for such action.

The *foundational* approach to energy justice is based on two cornerstones, the unique characteristics of energy as a good and insights from the applicable justice schools. These are used to create a philosophical foundation of energy justice and effectively define it through two principles, prohibitive and affirmative. Sovacool et al (2014) begin by identifying three key forms of justice – distributive, procedural and cosmopolitan – and proceed with four sequential assumptions (as depicted in Figure 2) (Jones et al, 2015), eventually arriving at the aforementioned two principles. In doing so they also enrich the analytical arsenal for applying the principles. This includes, for example, identifying the nature and contextualising of

Figure 2: Assumptions and principles of energy justice

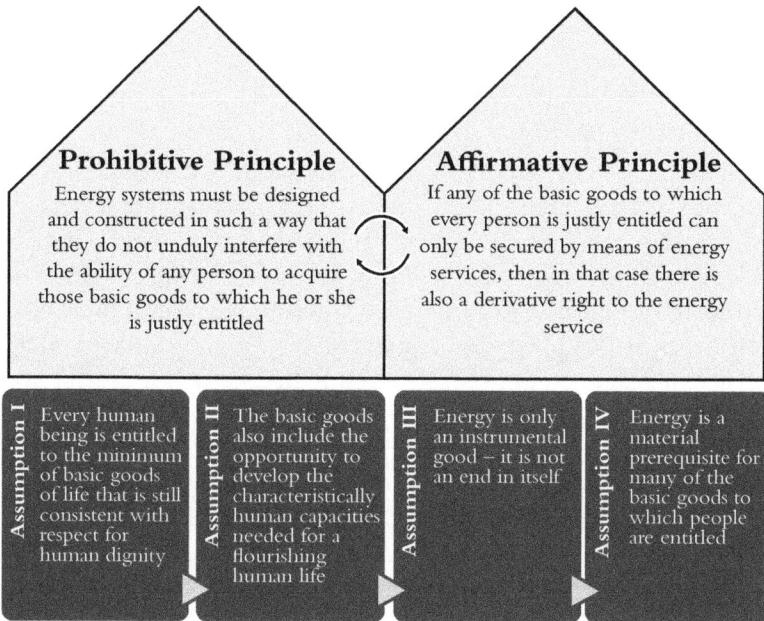

Source: Sovacool et al (2014, 2019) © Kathleen E. Halvorsen, Chelsea Schelly, Robert M. Handler, Erin C. Pischke and Jessie L. Knowlton 2019. Reproduced with permission of the Licensor through PLSclear.

one's response to energy injustice, as well as determining the ontology of basic goods (Jones et al, 2015). The interplay between the prohibitive and affirmative principles is a balancing act between an obligation to deliver energy services and the conditions under which the services need to be delivered. Unlike the system approach, the foundational approach also articulates why people have the right to modern energy services, as well as the extent of this right.

The centrepiece of the analysis to which Sovacool et al return repeatedly is the unique role that energy plays in the global economy and modern society. Energy is a prerequisite to many if not most goods, including basic ones. Thus, energy, instead of simply being means to other goods, instead dominates them. This domination transcends countries, communities and individual minds. It is used as a geopolitical weapon and a reason to risk the future of humankind. Policy discourse around energy is almost exclusively centred on energy commodities – barrels of oil, gallons of gasoline, megawatts of electricity and tonnes of coal – and not what all these sources of primary and secondary energy need to achieve. However, energy is not an end in itself. As Jones et al (2015, p 160) put it, 'the use of energy must be determined by the human ends it serves (rather than these ends being distorted to fit the technical imperatives of the energy system), and these ends must be consistent with respect for the equal dignity of human beings'.

Debates and key developments

Procedural energy justice and capabilities

Procedural justice is a key concern for individuals and communities who are affected by changes in energy systems. This is a shared point of concern for just transitions research (Chapter 15). Affected communities have been found to be routinely excluded from decision-making processes with regard to the siting of energy infrastructure in and around their vicinity (McCauley et al, 2013). Spaces of undue process are already well established in environmental and climate justice research. The unjust distribution of power plants or waste facilities is directly correlated with an ineffective or even absent process for including community organisations. Higginbotham et al (2010) revealed that residents in the Upper Hunter, Australia, were routinely blocked from inputting crucial scientific data on air pollution as part of their protest against the state's promotion of coal production in the area. More recently, similar research has emphasised the lack of procedural mechanisms for including opposition and supportive voices for shale gas in the UK (Cotton et al, 2014). This space of injustice is therefore characterised by non-inclusion in crucial decisions.

Research in procedural energy justice has been dominated by multiple case studies of wind energy. This reflects the broader changes in global energy systems (as well as a more Anglo-American dominance of literature as elaborated later in this chapter) towards investing in renewable energy sources. This has moved the debate on fair process from simply inclusion itself, towards reflections upon who is seeking to include and when this takes place. Warren and McFadyen (2010) demonstrated that local ownership of community wind farms has a greater chance to be accepted and incurs fewer instances of injustice. Ottinger et al (2014) find, in contrast, that greater state involvement can lead to less opposition in the US. Outside the EU and US examples, feelings of injustice on renewable energy are driven by the lack of informal recognition or appreciation of local livelihoods that are destroyed by some energy efficient projects (Yenneti and Day, 2015). Spaces of unfair process in emerging energy systems are therefore more complex, contextual and time-sensitive.

The decentralised nature of renewable energy systems requires a new approach to including affected communities in infrastructural decision-making. In the examples raised here, the community is viewed as detached from its energy system, at least in terms of production and associated processes. Originating from Sen (1999), the Capabilities Approach sheds light on not only the basic desire to access energy but also the wide range of capabilities that energy provides. Unconventional energy systems which do not require major infrastructure offer the potential for a much freer engagement for the traditionally understood consumer. Parag and Sovacool (2016) support this observation by suggesting that electricity markets are currently undergoing a process of redesign to deal with unconventional energy systems and smaller scale renewable systems.

Adjacent concepts: fuel poverty, energy poverty and energy vulnerability

Historically, fuel poverty is defined as the need to spend above 10 per cent of a household's income on energy. It is a practical action-oriented concept that exposes the structural unfairness of income poverty related to basic energy services. Fuel poverty preceded energy justice and motivated some energy scholars to think about structural unfairness in the distribution of energy services more broadly. Thus, it falls firmly under the umbrella of energy justice. In a study between 2010 and 2013, Sovacool (2015) found that the warm fronts programme in England significantly reduced the number of fuel poor British homes by providing energy efficiency upgrades. It led to a reduction in greenhouse gas emissions as well as an average annual income addition of £1,894 per household. Similar research has emphasised the need to investigate such spaces of fuel poverty due to higher levels of health issues (Lacroix and Chaton, 2015), resulting from damp or cold housing (Shortt and Rugkasa, 2007; Dear and McMichael, 2011) as well as inadequate air conditioning (Teller-Elsberg et al, 2016).

The concept of fuel poverty has been broadened by the development of energy poverty (Bouzarovski and Herrero, 2016). It relates to the injustices felt not only by those spending under the 10 per cent of annual income. Similarly to the concept of fuel poverty, energy poverty exposes structural unfairness related to income and wealth disparities in the context of the provision of energy services. It also belongs under the wide umbrella of energy justice as it connects particularly well with the aforementioned affirmative principle. It was instigated by the observation that fuel poverty is not as applicable outside developed nations. Energy poverty brings our attention to a much more absolute understanding of energy access. García-Ochoa and Graizbord (2010, p 40) define energy poverty, in relation to Mexico, as an agenda which seeks to reveal the 'deprivation of energy services linked to satisfying basic human needs'. The focus of responsibility is placed upon the providers of electricity, heating and transportation fuels.

Fuel poverty is also defined by misrecognition or exposing the understudied consequences of distribution which is often referred to as post-distributional conceptualisations (Walker and Day, 2012). A UK-based study found that poverty and lack of access to energy directly correlated with ill health among older people (de Vries and Blane, 2013). Research in energy poverty has attempted to move beyond post-distributional justice issues by considering procedural concerns also, for example in post-communist states of eastern and central Europe (Bouzarovski and Herrero, 2016).

From these origins, research in energy vulnerability (Fernández-González and Moreno, 2015; Middlemiss and Gillard, 2015; Cauvain and Bouzarovski, 2016; Bouzarovski et al, 2017a) emerged directly from the inclusion of post-distribution research (Bulkeley et al, 2014). Spaces of vulnerability are identified as the direct consequences of distributional injustices.

Responsibility in energy justice

Energy justice scholars have consistently argued that 'we all' (from government and business to citizens and academics) have a responsibility to ensure that energy justice is achieved (Heffron and McCauley, 2014; 2017; Reames, 2016; Sovacool et al, 2016; Munro et al, 2017). This understanding of responsibility builds directly upon the works of Iris Marion Young (2004; 2006; 2011). Young recognises that a shift in models of responsibility is required in order to respond to the major questions that society faces, such as climate change or making the global energy system more sustainable. She referred to this as the model of social connectedness, whereby individuals adopt senses of responsibility that go beyond their immediate context of family or even local community.

The trajectory of energy systems reinforces the argument that scholars need to adopt a broad understanding of responsibility. As the global energy system moves away from fossil fuels, alternative fuel systems are inherently more decentralised. The decentralisation of energy systems means that individuals and householders may decide to assume responsibilities for their own energy provision, as well as for others (Capaccioli et al, 2017). Their position within the energy system is not restricted to that of the end user. Recent studies (Ritzer et al, 2012; Parag and Sovacool, 2016) have focused on the notion of a prosumer, meaning a consumer that also produces for its own energy needs. Damgaard et al (2017) revealed in their study of biofuels in Nepal that individuals adopted a greater sense of responsibility in producing and consuming energy when they understood how their biofuel energy system worked, and that they had to maintain it.

No 'good' energy?

The global energy system must decarbonise to ensure sustainable long-term clean sources of electricity, heating and transport. The electricity sector has experienced the most improvement in this regard with the development of a wide range of sources (IEA, 2016). Renewable and low-carbon electricity technologies can, first, exude similar injustices as dirty fuels. The establishment of large-scale onshore and offshore wind and solar farms has placed communities in opposition to developers (Urpelainen, 2016; Yenneti and Day, 2016). Pepermans and Loots (2013) find that current siting processes in Flanders reinforced disagreements between communities and companies in a similar fashion to coal power stations. Bailey (2016) argues in a similar vein that national governments have exploited rural communities through the renewable agenda leading to the expansion of related infrastructures. From this perspective, clean fuels can exacerbate feelings of injustice, instigated by large-scale fossil fuel developments in the 1970s, through multiplying infrastructures in close proximity to communities.

The absence of health implications from air pollution commonly found in relation to fossil fuels does not necessarily translate into higher senses of justice. The comparatively large size of individual fossil fuel production infrastructure such as a power plant or waste facility contrast with micro- and medium-sized renewable sources (albeit large-scale wind farms often cover more space). This means that energy infrastructures are multiplied throughout a given region or nation (Liljenfeldt and Pettersson, 2017). In the case of wind turbines, this has marked a shift from urban-based concerns around justice towards rural communities (Malin, 2015). The high load needed for older and more established fossil fuel power plants require urban or semi-urban localisations (this is less true for newer power plants often located outside urban areas). This is not the case with wind or solar. The ability of a consumer to purchase and use micro infrastructure has moved energy towards the household level where energy can be both produced and consumed (Reid et al, 2009). Yet the size, location and scale matter, often more than the type of energy facility.

Connections with other applied theories
Spatial and intergenerational justice

Energy justice as a concept includes like-minded spatial conceptualisations such as proximity, due process and misrecognition. In terms of spatial approaches, energy justice has recently added a fourth dimension of restoration justice (Heffron and McCauley, 2017). The global energy system is a highly complex network of production, distribution, transmission and waste infrastructure designed to provide energy to end users. Traditionally, environmental justice would focus on the negative implications of energy and non-energy generating or waste-related activities for local populations (Tayarani et al, 2016). US-based research found that there was a high propensity of local, national and supranational organisations to locate these infrastructures within ethnic minority or socially deprived communities (Macias, 2016). Geographical literature in this area concentrates on revealing the place-specific nature of injustices, or explores the spatial tactics used by opposing activists (Houston, 2013). Proximity has therefore represented a key concern for such researchers. They did not, however, reflect explicitly on the role of energy. Energy justice provides an opportunity to fill this gap. The rise of new 'clean' energy technologies offers new spaces of proximity.

Energy production is the stereotypical focus when considering distributive justice implications. The standard examples often come from what is termed 'dirty fuels' such as oil (Orta-Martinez and Finer, 2010), especially coal (Higginbotham et al, 2010) or even gas (Nevins, 2004). These energy systems inevitably involve the construction of large centralised industrial plants to convert these sources of primary energy into secondary energy sources such as electricity and gasoline. During the 1970s, infrastructural developments took place throughout the US and Europe as the oil boom took place. This gave rise to justice-based research in these areas (Taylor, 2000). Today, the industrial fossil fuel system is either being updated or maintained in developed nations, while many emerging economies are currently, or recently, adopting large-scale fossil fuel systems. China, and especially India, are classic examples (Liu et al, 2014). From this perspective, new proximities to energy infrastructure emerge in developing world contexts where populations can be more vulnerable.

Nigeria is an example where oil has been the driver of the national political economy (Glazebrook and Kola-Olusanya, 2011). It has also fuelled conflict, as well as embedded logics of 'capital and care' (Maiangwa and Agbiboa, 2013). Proximity to oil fields leads to 'logics of capital' that is largely driven by militant male youths, whereas 'logics of care' are more closely associated with notions of justice driven by women. This demonstrates that injustice in energy does not always originate from the location of infrastructure in a given community as it can often be contextual, leading to variable outcomes for certain groups of society. The development of gas reserves in the Russian Arctic presents an example whereby energy is a more direct driver of injustice (MacDougall, 2001; McCauley et al, 2016). The decision of multinational energy companies to drill in the Yamal Peninsula has directly resulted in health implications for both the local community and foreign workers (Silin, 2015). The emergence of fracking has equally inspired opposition movements against shale gas in both the UK and US (Cotton et al, 2014; Eisenberg, 2015). Similar research has pointed to potential health implications of being located within the vicinity of the producing infrastructure (Crowe et al, 2015). In both cases, proximity has indeed resulted in social opposition and new feelings of injustice.

Intergenerational justice has also been at the heart of the climate mitigation policy discourse since the 2000s. It was spurred by the release of the 2006 *Stern Review on the Economics of Climate Change* (Stern, 2007) and, more specifically, the discount rate its authors adopted to value future damages from climate change. In essence, the discount rate is centred on the weight given to the welfare of future generations compared to the welfare of present generations. While the early discourse focused on the discount rate itself, energy and climate justice scholars added a dimension to the discourse by questioning the right of present generations to saddle future generations with a catastrophic debt. Sovacool et al (2014) argue that from the perspective of intergenerational justice, discounting the future impacts of climate change is nothing more than a ruse that serves to hide a terrifying indifference to the assured suffering and demise of millions of people yet to be born. Factoring in current, likely and possible impacts of climate change on agricultural land, freshwater resources and fisheries, the belief that future generations will have the collective wealth to deal with these impacts is problematic if not delusional. Moreover, it is not just the debt of climate change impacts that is at issue, if the configuration of current energy systems is not scrutinised temporally, future generations will be saddled with an obsolete energy infrastructure incapable of providing even basic energy services.

Environmental and climate justice

Although environmental and climate justice target the energy sector in some ways, neither can sufficiently encapsulate emerging questions around equity and fairness with regard to current and future energy systems. Environmental and climate justice are well-established literature bases in multiple disciplines, for example, geography, sociology and environmental studies. Neither can sufficiently encapsulate emerging questions around equity and fairness with regard to current and future energy systems. Environmental justice has been a successful tool for activists (Schlosberg, 2004; Houston, 2013). Its origins are closely related to social opposition against the siting of hazardous waste in the US. Studies emerged in academia as opportunities to reflect upon the ways in which these injustices were resisted (Taylor, 2000). Similar research has also emerged outside the US, often focusing on resistance movements including in Africa (Ako, 2009), Europe (Slater and Pedersen, 2009), South America (Urkidi and Walter, 2011) and Southeast Asia. Early research in this area reveals the distributional injustices with regards to where environmental burdens are sited (Taylor, 2000). It sheds light on how companies and governments sited harmful infrastructure through planning processes in areas of social deprivation or near ethnic minority communities (Shrader-Frechette, 1996). More recent literature has offered insight into decision-making processes which have been referred to as investigations of procedural justice (Hricko et al, 2014). Scholars realised in this way that the process of locating infrastructure was equally important as the final outcome.

The focus in environmental justice research is therefore positioned at the intersection between social concern and environmental impacts. It is equally valid for other forms of research where the emphasis is placed outside the environment. Climate justice emerged directly from this literature and associated conceptual frameworks (Bulkeley et al, 2013; Harris et al, 2013; Olawuyi, 2016). The focus of resistance is placed directly upon a much larger concern than individual environmental impacts (Russell, 2015). Climate change is presented as an overriding meta concern where social justice is juxtaposed with international climate negotiations (Lyster, 2017), their implementation (Mathur et al, 2014) and the local consequences of rapid changes in climate (Bulkeley et al, 2014). Conceptually, this agenda brought a new spatial dimension to academic research in the form of misrecognition, albeit absent from some climate justice literature in geography (Fisher, 2015). It encouraged us to consider who is missing from our policies or decisions in response to climate change (see Chapter 10).

Applying energy justice in the social sciences

As noted in this chapter, the concept of energy justice emerged in the social sciences and has been developed largely by empirical scholars. Its rapid ascent and adoption have already benefited energy research and there is no dire need to tighten theoretical conventions just yet. Theoretical accounts of justice might restrict energy justice researchers in activism (and more generally) pigeonhole them into predetermined western conceptions of justice (Barnett, 2010). Attention should be drawn to where and when injustice is felt and experienced, in support of Hobson (2006), justice-based activism research must diversify its understanding of where injustice can be found in multiple contexts. Justice, in this regard, is pluralist.

Reed and George (2011, p 839) comment that 'researchers are cautioned that the long observed disconnect between theory and practice in the field of environmental justice may be exacerbated should academics become more concerned with theoretical refinement over progressive, practical, and possible change'. The theorisation of justice seeks to expose ideal endpoints (and more recently processes) from various (usually western) philosophical traditions. In a similar vein, Schlosberg (2013) argues that justice theorists need to be pluralist in accepting a range of understandings of 'good'. The first step in this direction is therefore the acknowledgement that the study of justice is pluralist. It is argued here that we need to explore the plurality of injustice too.

Martin et al (2014, p 2) acknowledge 'that justice poses considerable conceptual challenges, not least because of the practical (if not intellectual) impossibility of reaching consensus'. Their conclusion bears a self-reflective unease; as they question the limitations of their own framing and methods, including the underlying logics of justice. This calls for acknowledgement, then, that justice is contextual. Walker (2009, p 622) comments, for example, that 'as we move from concern to concern and from context to context, we can expect shifts in both the spatial relations that are seen to be significant and in the nature of justice claims being made'. The expansion in the theorisation of justice as a concept must be answered with a similar response in our empirical understanding of energy justice and the injustices it entails. As Barnett comments in support of Sen (2011):

> Rather than thinking of philosophy as a place to visit in order to find idealised models of justice or radically new ontologies, we would do well to notice that there is an identifiable shift among moral and political philosophers towards starting from more worldly, intuitive understandings of injustice, indignation, and harm, and building up from there. (Barnett, 2010, p 252)

Energy justice is a foundational concept for social scientists to investigate the ethics, morality and values behind energy decision-making, the negative and positive outcomes thereof, and the causal links and gaps between the decisions and impacts. Energy justice is often best identified and analysed by examining energy injustice. However, energy inequities usually span many dimensions making it difficult to identify and classify their exact origins. Yet it is challenging to solve an energy injustice problem without knowing *how* it impacts people and the environment and *where* the impacts are felt the most. Having a typology helps account for the complexity of energy justice problems while designing solutions that target the causes and not symptoms.

Sovacool et al (2014) offer such a typology and Sidortsov and Sovacool (2015) apply it in the context of energy development in the Arctic. The typology is centred on five dimensions – temporal, economic, sociopolitical, geographic and technological – and was developed based on the prevalent energy injustices that the scholars identified as part of their work. The temporal dimension groups injustices arising from harmful legacies that are passed to future generations. The economic dimension puts a spotlight on economic inequalities, inequitable distribution of energy services, energy and fuel poverty and drudgery, energy price volatility, and the economic impacts of resource depletion. The sociopolitical dimension highlights conflicts over energy resources, resource curse, social marginalisation, corruption, often leading to the erosion of democratic institutions, and human rights abuses. The geographic dimension helps to identify injustices related to the unequal distribution of risks, impacts and benefits of energy development such as the creation of sacrifice zones, community displacement and climate refugees. The technological dimension refers to injustices that are embedded in the design of certain energy technologies: reliability, safety, path dependence, vulnerability and inefficiency.

The Five Dimensions of Energy Justice framework is just one example of the array of impressive analytical tools that energy justice scholars have amassed. These tools cover most, if not all, social science disciplines and can be used for descriptive, evaluative and prescriptive purposes. Energy justice remains a social science concept throughout that will continue to serve researchers, practitioners and activists well.

Conclusion

Barely a decade old, energy justice has already emerged as a foundational concept in energy studies. Inherently transdisciplinary, energy justice transcends academia and is emerging as an analytical tool employed by activists and practitioners alike. Its strong empirical foundation safeguards it against the domination of a single normative justice theory. Rather, energy justice scholars draw upon different traditions of justice, often employing distributive, procedural, recognition and restorative forms of justice to develop the concept. There is no unity among scholars in defining energy justice, which is a good thing. Energy justice discourse remains fresh and stimulating with various frameworks borrowing from each other and not trying to prove each other wrong. Even the two seemingly divergent approaches to energy justice, system and foundational, can and do work together as each provides a different perspective on an energy justice problem at hand. Energy justice works well with several adjacent concepts such as energy and fuel poverty and energy vulnerability. Whereas these concepts enable a researcher to zoom in on a specific issue, affordability of heating for example, energy justice provides an overarching framework for placing and assessing this issue within the ethics of the energy cycle or energy system.

What makes energy justice a distinct applied theory of justice is the special nature of energy as a good. In theory, a means of achieving the end, moving people from point A to point B, for example, it tends to dominate the end, in this instance, chaining it to the global oil market. Thus, other relevant applied theories, environmental, climate, spatial and intergenerational justice, only address some injustices created by energy systems. These injustices often fall into the category of burdens created by the production, transportation and use of energy, leaving the services that energy systems must deliver unaddressed. However, this makes these applied justice concepts perfect complementary tools in the hands of energy researchers and practitioners to transition the world towards a sustainable and equitable future.

Notes

[1] For the purposes of this chapter, we refer to energy as primary energy and secondary energy. The International Energy Agency defines primary energy that the energy stored in natural resources that has not undergone any anthropogenic conversion. Primary energy that has been modified for a particular use, refined into petroleum and converted into electricity but not delivered to final users, is referred to as secondary energy (Intergovernmental Panel on Climate Change and Edenhofer, 2014).

2 According to the Intergovernmental Panel on Climate Change, 'the energy system comprises all components related to the production, conversion, delivery, and use of energy' (Intergovernmental Panel on Climate Change and Edenhofer, 2014, p 1261).

References

Ako, R.T. (2009) Nigeria's and use act: An anti-thesis to environmental justice. *Journal of African Law*, 53(2), 289–304.

Bailey, I. (2016) Renewable energy, neoliberal governance and the tragedy of the Cornish commons. *Area*, 48(1), 119–121.

Barnett, C. (2010) Geography and ethics: Justice unbound. *Progress in Human Geography*, 35(2), 246–255.

Barrett, S. (2012) The necessity of a multiscalar analysis of climate justice. *Progress in Human Geography*, 37(2), 215–233.

Bickerstaff, K. and Agyeman, J. (2009) Assembling justice spaces: The scalar politics of environmental justice in north-east England. *Antipode*, 41(4), 781–806.

Bouzarovski, S. and Herrero, S.T. (2016) Geographies of injustice: The socio-spatial determinants of energy poverty in Poland, the Czech Republic and Hungary. *Post Communist Economies*, 29(1), 27–50.

Bouzarovski, S., Herrero, S.T. and Petrova, S. (2017a) Energy vulnerability trends and factors in Hungary. *Energie und Soziale Ungleichheit*, 455–474.

Bouzarovski, S., Herrero, S. and Petrova, S. (2017b) Multiple transformations: Theorizing energy vulnerability as a socio-spatial phenomenon. *Geografiska Annaler Series B: Human Geography*, 99(1), 20–41.

Bulkeley, H., Carmin, J. and Castán Broto, V. (2013) Climate justice and global cities: Mapping the emerging discourses. *Global Environmental Change*, 23(5), 914–925.

Bulkeley, H., Edwards, G.A.S. and Fuller, S. (2014) Contesting climate justice in the city: Examining politics and practice in urban climate change experiments. *Global Environmental Change*, 25, 31–40.

Capaccioli, A., Poderi, G. and Bettega, M. (2017) Exploring participatory energy budgeting as a policy instrument to foster energy justice. *Energy Policy*, 107, 621–630.

Cauvain, J. and Bouzarovski, S. (2016) Energy vulnerability in multiple occupancy housing: A problem that policy forgot. *People, Place & Policy Online*, 10(1), 88–106.

Cotton, M., Rattle, I. and Van Alstine, J. (2014) Shale gas policy in the United Kingdom: An argumentative discourse analysis. *Energy Policy*, 73, 427–438.

Crowe, J., Silva, T. and Ceresola, R.G. (2015) Differences in public perceptions and leaders' perceptions on hydraulic fracturing and shale development. *Sociological Perspectives*, 58(3), 441–463.

Damgaard, C., McCauley, D. and Long, J. (2017) Assessing the energy justice implications of bioenergy development in Nepal. *Energy, Sustainability and Society*, 7, 8. https://doi.org/10.1186/s13705-017-0111-6

Dear, K.B. and McMichael, A.J. (2011) The health impacts of cold homes and fuel poverty. Report. https://www.instituteofhealthequity.org/resources-reports/the-health-impacts-of-cold-homes-and-fuel-poverty

de Vries, R. and Blane, D. (2013) Fuel poverty and the health of older people: The role of local climate. *Journal of Public Health*, 35(3), 361–366.

Eisenberg, A.M. (2015) Beyond science and hysteria: Reality and perceptions of environmental justice concerns surrounding Marcellus and Utica shale gas development. *University of Pittsburgh Law Review*, 77(2), 183–234.

Fernández-González, P. and Moreno, B. (2015) Analyzing driving forces behind changes in energy vulnerability of Spanish electricity generation through a Divisia index-based method. *Energy Conversion and Management*, 92, 459–468.

Fisher, S. (2015) The emerging geographies of climate justice. *Geographical Journal*, 181(1), 73–82.

García-Ochoa, R. and Graizbord, B. (2016) Privation of energy services in Mexican households: An alternative measure of energy poverty. *Energy Research & Social Science*, 18, 36–49.

Glazebrook, T. and Kola-Olusanya, A. (2011) Justice, conflict, capital, and care: Oil in the Niger Delta. *Environmental Ethics*, 33(2), 163–184.

Guruswamy, L. (2010) Energy justice and sustainable development. *Colorado Journal of International Environmental Law and Policy*, 21, 231.

Guruswamy, L. (2015) Global energy justice. In L. Guruswamy (ed), *International Energy and Poverty: The Emerging Contours*. London: Routledge, pp 55–67.

Guruswamy, L. (2016) *Global Energy Justice: Law and Policy*. St. Paul, MN: West Academic Publishing.

Hall, S. (2013) Energy justice and ethical consumption: Comparison, synthesis and lesson drawing. *Local Environment*, 18, 422–437. 10.1080/13549839. 2012.748730.

Harris, P., Chow, A. and Karlsson, R. (2013) China and climate justice: Moving beyond statism. *International Environmental Agreements-Politics Law and Economics*, 13(3), 291–305.

Heffron, R.J. and McCauley, D. (2014) Achieving sustainable supply chains through energy justice. *Applied Energy*, 123, 435–437.

Heffron, R.J. and McCauley, D. (2017) The concept of energy justice across the disciplines. *Energy Policy*, 105, 658–667.

Higginbotham, N., Freeman, S. and Connor, L. (2010) Environmental injustice and air pollution in coal affected communities, Hunter Valley, Australia. *Health & Place*, 16(2), 259–266.

Hiteva, R. and Sovacool, B. (2017) Harnessing social innovation for energy justice: A business model perspective. *Energy Policy*, 107, 631–639.

Hobson, K. (2006) Enacting environmental justice in Singapore: Performative justice and the Green Volunteer Network. *Geoforum*, 37(5), 671–681.

Houston, D. (2013) Environmental justice storytelling: Angels and isotopes at Yucca Mountain, Nevada. *Antipode*, 45(2), 417–435.

Hricko, A., Rowland, G. and Eckel, S. (2014) Global trade, local impacts: Lessons from California on health impacts and environmental justice concerns for residents living near freight rail yards. *International Journal of Environmental Research and Public Health*, 11(2), 1914–1941.

International Energy Agency (IEA) (2016) *World Energy Statistics 2016*. Paris: IEA. https://www.iea.org/reports/world-energy-outlook-2016

Intergovernmental Panel on Climate Change and Edenhofer, O. (eds) (2014) *Climate Change 2014: Mitigation of Climate Change: Working Group III contribution to the Fifth Assessment Report of the Intergovernmental Panel on Climate Change*. Cambridge, UK: Cambridge University Press.

Jenkins, K., Heffron, R.J. and McCauley, D. (2016) The political economy of energy justice: A nuclear energy perspective. In T. Van de Graaf, B. Sovacool and A. Ghosh (eds), *Palgrave Handbook of the International Political Economy of Energy*. Basingstoke, UK: Palgrave, p 661–682.

Jones, B.R., Sovacool, B.K., Sidortsov, R.V. and Center for Environmental Philosophy, The University of North Texas (2015) Making the ethical and philosophical case for 'energy justice'. *Environmental Ethics*, 37(2), 145–168.

Lacroix, E. and Chaton, C. (2015) Fuel poverty as a major determinant of perceived health: The case of France. *Public Health*, 129(5), 517–524.

Liljenfeldt, J. and Pettersson, Ö. (2017) Distributional justice in Swedish wind power development: An odds ratio analysis of windmill localization and local residents' socio-economic characteristics. *Energy Policy*, 105, 648–657.

Liu, L., Liu, J. and Zhang, Z. (2014) Environmental justice and sustainability impact assessment: In search of solutions to ethnic conflicts caused by coal mining in inner Mongolia, China. *Sustainability*, 6(12), 8756–8774.

Lyster, R. (2017) Climate justice, adaptation and the Paris Agreement: A recipe for disasters? *Environmental Politics*, 26(3), 1–21.

MacDougall, J.C. (2001) Access to justice for deaf Inuit in Nunavut: The role of 'Inuit Sign Language'. *Canadian Psychology-Psychologie Canadienne*, 42(1), 61–73.

Macias, T. (2016) Environmental risk perception among race and ethnic groups in the United States. *Ethnicities*, 16(1), 111–129.

Maiangwa, B. and Agbiboa, D.E. (2013) Oil multinational corporations, environmental irresponsibility and turbulent peace in the Niger Delta. *Africa Spectrum*, 48(2), 71–83.

Malin, S. (2015) Conclusions and solutions: Social sustainability and localized energy justice. In S. Malin (ed), *The Price of Nuclear Power: Uranium Communities and Environmental Justice*. New Brunswick, NJ: Rutgers University Press, pp. 148–160.

Martin, A., Gross-Camp, N. and Kebede, B. (2014) Whose environmental justice? Exploring local and global perspectives in a payments for ecosystem services scheme in Rwanda. *Geoforum*, 54, 167–177.

Mathur, V.N., Afionis, S. and Paavola, J. (2014) Experiences of host communities with carbon market projects: Towards multi-level climate justice. *Climate Policy*, 14(1), 42–62.

McCauley, D., Heffron, R. and Stephan, H. (2013) Advancing energy justice: The triumvirate of tenets. *International Energy Law Review*, 32(3), 107–111.

McCauley, D., Heffron, R. and Pavlenko, M. (2016) Energy justice in the Arctic: Implications for energy infrastructural development in the Arctic. *Energy Research & Social Science*, 16, 141–146.

Middlemiss, L. and Gillard, R. (2015) Fuel poverty from the bottom-up: Characterising household energy vulnerability through the lived experience of the fuel poor. *Energy Research and Social Science*, 6, 146–154.

Munro, P., van der Horst, G. and Healy, S. (2017) Energy justice for all? Rethinking sustainable development goal 7 through struggles over traditional energy practices in Sierra Leone. *Energy Policy*, 105, 635–641.

Nevins, J. (2004) Contesting the boundaries of international justice: State countermapping and offshore resource struggles between east Timor and Australia. *Economic Geography*, 80(1), 1–22.

Olawuyi, D.S. (2016) Climate justice and corporate responsibility: Taking human rights seriously in climate actions and projects. *Journal of Energy & Natural Resources Law*, 34(1), 27–44.

Orta-Martinez, M. and Finer, M. (2010) Oil frontiers and indigenous resistance in the Peruvian Amazon. *Ecological Economics*, 70(2), 207–218.

Ottinger, G., Hargrave, T.J. and Hopson, E. (2014) Procedural justice in wind facility siting: Recommendations for state-led siting processes. *Energy Policy*, 65, 662–669.

Parag, Y. and Sovacool, B. (2016) Electricity market design for the prosumer era. *Nature Energy*, 1, 1–12.

Pepermans, Y. and Loots, I. (2013) Wind farm struggles in Flanders fields: A sociological perspective. *Energy Policy*, 59, 321–328.

Reames, T. (2016) Targeting energy justice: Exploring spatial, racial/ethnic and socioeconomic disparities in urban residential heating energy efficiency. *Energy Policy*, 97, 549–558.

Reed, M.G. and George, C. (2011) Where in the world is environmental justice? *Progress in Human Geography*, 35(6), 835–842.

Reid, L., Sutton, P. and Hunter, C. (2009) Theorizing the meso level: The household as a crucible of pro-environmental behaviour. *Progress in Human Geography*, 34(3), 309–327.

Ritzer, G., Dean, P. and Jurgenson, N. (2012) The coming of age of the Prosumer. *American Behavioral Scientist*, 56(4), 379–398.

Russell, B. (2015) Beyond activism/academia: Militant research and the radical climate and climate justice movement(s). *Area*, 47(3), 222–229.

Schlosberg, D. (2004) Reconceiving environmental justice: Global movements and political theories. *Environmental Politics*, 13(3), 517–540.

Schlosberg, D. (2013) Theorising environmental justice: The expanding sphere of a discourse. *Environmental Politics*, 22(1), 37–55.

Sen, A. (1999) *Development as Freedom*. New York: Oxford University Press.

Sen, A. (2011) *The Idea of Justice*. Cambridge, MA: Harvard University Press.

Shortt, N. and Rugkasa, J. (2007) 'The walls were so damp and cold' fuel poverty and ill health in Northern Ireland: Results from a housing intervention. *Health & Place*, 13(1), 99–110.

Shrader-Frechette, K. (1996) Environmental justice and Native Americans: The Mescalero Apache and monitored retrievable storage. *Natural Resources Journal*, 36(4), 943–954.

Sidortsov, R.V., Heffron, R.J., Mose, T.M., Schelly, C. and Tarekegne, B. (2019) In search of common ground: energy justice perspectives in global fossil fuel extraction. In K.E. Halvorsen, C. Schelly, R.M. Handler, E.C. Pischke and J.L. Knowlton (eds) *A Research Agenda for Environmental Management*. Cheltenham UK: Edward Elgar.

Sidortsov, R.V. and Sovacool, B.K. (2015) Left out in the cold: Energy justice and Arctic energy research. *Journal of Environmental Studies and Sciences*, 5(3), 302–307.

Sidortsov, R.V., Heffron, R.J., Mose Moya, T., Schelly, C. and Tarekegne, B. (2019) In search of common ground: Energy justice perspectives in global fossil fuel extraction. In K. Halvorsen, C. Schelly, R. Handler, E. Pischke and J. Knowlton (eds), *A Research Agenda for Environmental Management*. Cheltenham, UK: Edward Elgar, pp 134–144.

Silin, A.N. (2015) Sociological aspects of rotational employment in the northern territories of Western Siberia. *Economic and Social Changes: Facts, Trends, Forecast*, 4(40), 109–123.

Slater, A.-M. and Pedersen, O.W. (2009) Environmental justice: Lessons on definition and delivery from Scotland. *Journal of Environmental Planning and Management*, 52(6), 797–812.

Sovacool, B.K. (2015) Fuel poverty, affordability, and energy justice in England: Policy insights from the Warm Front Program. *Energy*, 93, 361–371.

Sovacool, B.K. and Dworkin, M.H. (2014) *Global Energy Justice: Problems, Principles, and Practices*. Cambridge, UK: Cambridge University Press.

Sovacool, B.K., Heffron, R.J. and McCauley, D. (2016) Energy decisions reframed as justice and ethical concerns. *Nature Energy*, 1, 16–24.

Sovacool, B.K., Sidortsov, R. and Jones, B. (2014) *Energy Security, Equality, and Justice*. London: Routledge.

Stern, N. (2007) *The Economics of Climate Change: The Stern Review*. Cambridge, UK: Cambridge University Press.

Tayarani, M., Poorfakhraei, A. and Nadafianshahamabadi, R. (2016) Evaluating unintended outcomes of regional smart-growth strategies: Environmental justice and public health concerns. *Transportation Research Part D-Transport and Environment*, 49, 280–290.

Taylor, D.E. (2000) The rise of the environmental justice paradigm: Injustice framing and the social construction of environmental discourses. *American Behavioral Scientist*, 43(4), 508–580.

Teller-Elsberg, J., Sovacool, B. and Smith, T. (2016) Fuel poverty, excess winter deaths, and energy costs in Vermont: Burdensome for whom? *Energy Policy*, 90, 81–91.

Urkidi, L. and Walter, M. (2011) Dimensions of environmental justice in anti-gold mining movements in Latin America. *Geoforum*, 42(6), 683–695.

Urpelainen, J. (2016) Energy poverty and perceptions of solar power in marginalized communities: Survey evidence from Uttar Pradesh, India. *Renewable Energy*, 85, 534–539.

Walker, G. (2009) Beyond distribution and proximity: Exploring the multiple spatialities of environmental justice. *Antipode*, 41(4), 614–636.

Walker, G. and Bulkeley, H. (2006) Geographies of environmental justice. *Geoforum*, 37(5), 655–659.

Walker, G. and Day, R. (2012) Fuel poverty as injustice: Integrating distribution, recognition and procedure in the struggle for affordable warmth. *Energy Policy*, 49, 69–75.

Warren, C.R. and McFadyen, M. (2010) Does community ownership affect public attitudes to wind energy? A case study from south-west Scotland. *Land Use Policy*, 27(2), 204–213.

Yenneti, K. and Day, R. (2015) Procedural (in)justice in the implementation of solar energy: The case of Charanaka solar park, Gujarat, India. *Energy Policy*, 86, 664–673.

Yenneti, K. and Day, R. (2016) Distributional justice in solar energy implementation in India: The case of Charanka solar park. *Journal of Rural Studies*, 46(1), 35–46.

Young, I.M. (2004) Responsibility and global labor justice. *Journal of Political Philosophy*, 12(4), 365–388.

Young, I.M. (2006) Responsibility and global justice: A social connection model. *Social Philosophy & Policy*, 23(1), 102–130.

Young, I.M. (2011) *Responsibility for Justice*. Oxford, UK: Oxford University Press.

12

Spatial Justice

Stephen Przybylinski

Introduction

The idea that justice or injustice takes place somewhere, that there is a geography to justice, is central to the notion of 'spatial justice'. Spatial justice is premised on the idea that justice or injustice has 'a consequential geography' to its formulation (Soja, 2010, p 1). While this may be so, the development of a specific *theory* of spatial justice has been anything but straightforward. For, just as the broader concept of justice has been widely interpreted, so too have there been ever changing positions on how spatial thinking matters for conceptualising justice. There is, then, no one definition of spatial justice, but, rather, a range of meanings that are realised through the application of the concept to specific phenomena. As this chapter will detail, while the concept of spatial justice has its origins in geographic theories of liberal, distributive justice, it has been substantially developed through Marxian, feminist, post-modern and post-structural approaches as well. Given the varying positions on justice within these traditions, spatial justice cannot be said to be a defined spatial theory of justice as much as it is an analytical framework for identifying situations where injustice arises.

Before analysing the development of the concept of spatial justice, I briefly introduce the connection between the concepts of space and scale. To understand how injustice takes place, an understanding of these key concepts is essential. Space is often conceptualised in at least three ways: space as 'absolute', 'relative' and 'relational'. *Absolute* space is generally understood as physical space that holds a fixed location and that is mappable. The city hall and the library, for example, are absolute spaces. *Relative* space is the perception of propinquity between absolute spaces based on different factors, for example, time or physical distance. For instance, the city hall stands *near* the library. Nearness is relative. I can be one hour away from the library when walking, but only 20 minutes away when biking. Finally, *relational* space is the space 'produced' by social relations. That is, relational space is space 'folded into social relations through practical activities' (Gregory, 2009, p 708). To keep with the previous example, city halls or libraries are more than simply containers where people and things are located *in*. They are also spaces made up of or produced through a wide range of social relations that give them various representational forms. The city hall is thus at once an absolute space as a physical object while simultaneously a set of relations. It is a social space, whereby employees go there to work as well as a space of representation for citizens and elected officials to make decisions about a given issue. When space is understood in this way, whereby a mix of social relations produce or shape space, we are provided a means of examining social relations through the spatial forms that these relations take. Importantly, these three categories are not exclusive. Space is absolute, relative and relational at the same time.

Inseparable from any understanding of space is the concept of scale. Discussions about scale have become theoretically complex within geography,[1] but to understand scale is to make sense of how space is relational. Most accept that scale is not ontologically pre-given, but, instead, submit that it is operationalised through the relations and processes producing space. In other words, there is not only one rigidly defined scale that exists when examining how space is produced. For instance, when the 'local' scale is referred to, the meaning of local is relative to the phenomena being studied. With that in mind, what is most important to consider is *how* scale is produced or constructed and *why* this analytical distinction matters. A 'politics of scale' is a means of conceptualisation that is sensitive to the ways in which ontological depictions of a defined or constant scale, for example, the nation state, is used to reinforce or disrupt certain relations in space. In this way, scale is as much a measure for bounding analyses as it is a concept used to explain certain socio-spatial processes.

The city or urban scale in particular has been a point of focus regarding justice in geography and urban planning scholarship. Scholars have assessed the 'justness' or 'injustice' of urban development, for example, by evaluating the development of the city through a lens of equity, democracy, diversity (Fainstein, 2010); by focusing on the role of urban institutions in advancing justice (Moroni, 2020); and by addressing the constraints of advancing urban justice within a liberal individualist framework more generally (Smith, 1995). Here, the scale of injustices examined matters for how and for whom space is produced. Yet, despite much critical scholarly attention to socio-spatial inequities, justice as a normative concept has received less analytical attention as it relates to spatial thinking (Przybylinski, 2022). This is not to say that spatial thinking has not contributed to justice-oriented scholarship at all. For, the concept of spatial justice has long been in development. The following section traces the concept's development before pointing to some connections between spatial thinking about justice with more normative theories of space which have not yet been developed through the concept of justice.

Spatial justice: the development of the concept

Originally published in 1973, David Harvey's (2009) book *Social Justice and the City* worked out the first explicitly geographic contribution to justice theorising. In the first half of *Social Justice*, Harvey followed the work of John Rawls, whose theory of 'justice as fairness' sought to collapse the inequities between groups by allocating resources to the least well-off (see Chapter 1). Harvey adapted this distributive notion of justice to that of 'territorial justice,' which he noted would require 'a form of spatial organization that maximizes the prospects of the least fortunate region' (2009, p 110). To find territorial justice, Harvey noted, would require 'a just distribution justly arrived at' (2009, p 98). Such a turn of phrase indicated where Harvey was heading conceptually. For, he would go on to argue that liberal distributive theories are limited in that that they do not make the distinction between distribution and *production*. Mainstream distributive theories assume that 'production and distribution are related to each other and that efficiency in the one is related to equity in the other' (Harvey, 2009, p 15). What distributive theories do not recognise, he argued, is that 'production *is* distribution and that efficiency *is* equity in distribution' (2009, p 15). In turning his attention towards the processes of urbanisation which facilitate the circulation of surplus value throughout the built environment, Harvey moved away from developing an explicitly territorial conception of justice, arguing instead that injustice in general derives from the ways in which capitalist production produces unequal social relations in space.

The spatial disciplines would not substantively examine conceptions of justice for nearly two decades after *Social Justice and the City*. It was then, in the early 1990s, geographers began moving beyond theorising justice simply as an issue of spatial distribution alone. Two rather different approaches to geographical inquiry into justice developed at this time. On the one hand, there was an explicit call for geographers to engage with the concept of social justice through moral and ethical approaches. David Smith was the leading figure calling for more explicit engagement with ethical and moral reasoning within geographical approaches to justice. Smith noted that geographers' preoccupation with distribution limited how geographical thinking informs an understanding of justice/injustice. He called for more ethical reflection and normative thinking in geography to push scholars to identify what constitutes a conception of the good life in order to better assess human well-being with the goal of better identifying how to reduce inequalities (2000, p 1156). Although Smith did not advance an explicit geographic theory of justice, his work stands out for expanding how justice can be uniquely informed through a geographical perspective. He reminds us that all humans need a place to be and, because of this, there are aspects of how geography relates to social justice that may illuminate 'the spatial variation of life chances as a dimension of inequality' (1994, p 296).

A second approach in this period followed the influential arguments of feminist philosopher Iris Marion Young (1990) whose ideas were revolutionary in shaping the conversation on justice within and outside of spatial disciplines (see Chapter 4). Among Young's most influential set of ideas was her assertion that justice cannot simply be a matter of (re)distribution alone, as mainstream liberal justice theories had made it, because of how distributive theories overlook other significant aspects of injustice. Rather, she argued that justice or injustice requires attention to the multiple oppressions and domination experienced by groups and how self-development and self-determination are constrained by institutions (1990, pp 35–68). Key for Young's theorising of justice was the need to balance the particularities of social *difference* while at the same time promoting universal values of self-determination and development. Young, among others, provided a fundamental advance in justice theorising which pushed scholars to articulate how space mattered to conceptions of justice beyond distributive (territorial) justice alone.

Influenced by Young, it was again David Harvey who advanced a theory of geographical difference. Key to such a geographic analysis was to explain how relations of oppression are produced through geographic relations. For Harvey, the context required to better understand matters of justice could be found by analysing the production of geographic space. Harvey (1992; 1996) thus added to Young's (1990, pp 48–63) five faces of oppression – exploitation, marginalisation, powerlessness, cultural imperialism and violence – to reinforce the importance of geography in producing difference and oppression in social relations. He added a sixth oppression: 'eco-generational', which recognised that 'the necessary ecological consequences of all social projects have impacts on future generations as well as upon distant peoples' (Harvey, 1992, p 600). To recognise this as a sixth oppression requires attention to the (geographic) consequences of human appropriation and transformation of the world around people in the making of their histories.

Although a means towards explaining geographical difference, Harvey's discursive addition to Young's categories of injustice was not intended as an explicit theory of spatial justice, however. Rather, his geo-historical materialist analysis promoted a geographically sensitive framework for evaluating injustice, one using geographic concepts of space and scale to evaluate relations of justice. Until the turn of the 21st century, therefore, there was no specific concept of spatial justice as such. This would change in the early 2000s, when geographers and those in other related disciplines began to develop upon ontological views of space which saw not only that the production of (or maldistribution of goods in) space created injustices, but that spatial relations themselves were constitutive of injustice.

The use of the term 'spatial justice' became popularised in the early 2000s and is typically associated with the post-modern geographer Edward Soja. Soja's work continues to be influential, inspiring scholars searching for deeper ontological explanations of spatiality. Heavily influenced by the work of the philosopher Henri Lefebvre, Soja's (1980; 1989) significant contribution to spatial thinking was developed in his notion of the 'socio-spatial dialectic'. The socio-spatial dialectic was intended to show how spatial relations themselves influenced social relations. For Soja (1980, p 211), the idea is that 'social and spatial relations are dialectically inter-reactive, inter-dependent ... social relations of production are both space-forming and space-contingent'. In reaction to the materialist analyses of Marxist geographers inspired by Harvey, for instance, Soja argued that the organisation of space was not merely an outcome of the capitalist mode of production but that spatial relations were a 'dialectically defined component of the general relations of production, relations which are simultaneously social and spatial' (1980, p 208). This idea would be paramount for and developed throughout the rest of his scholarship.

Although popularised much later, the first mention of the term spatial justice was made in the early 1980s. It was then that Gordon Pirie, a South African geographer, pondered whether a theory of spatial justice could help geographers move beyond the constraints of territorial distributive justice, an issue Harvey had addressed in *Social Justice and the City*. With this in mind, Pirie argued that if the concept of spatial justice was to be more than simply a way of indicating 'justice in space' (1983, p 469), then there was a need to develop an explicitly spatial conception of justice. Developing such a theory, Pirie thought, could offer a useful framework of analysis for evaluating social justice, as a means of 'spatial judgement', given that there are so many ways of understanding social justice. Pirie would not go on to develop a theory of justice. Nonetheless, he acknowledged the limitations of distributive notions of territorial justice and provided a language which foreshadowed the coming dialectical analysis of space and social justice.

Two decades later, taking up where Pirie left off, Mustafa Dikeç advanced a notion of justice which sought to illustrate the dynamism of socio-spatial relations which were less present in the distributive concerns of territorial notions of justice. Dikeç's (2001, pp 1787–1788) interest in spatial justice, following Soja's reconceptualisation of space, was that 'the very production of space ... not only manifests various forms of injustice, but actually produces *and* reproduces them (thereby maintaining established social relations of domination and oppression)' (emphasis in original). Dikeç's particular emphasis was on the dialectical relationship of space with injustice. It was not simply that the outcome of the social production of space created injustices, but that injustice could be discerned from spatiality itself, or those spatial relations which eliminated 'the possibilities for the formation of political responses' to various issues (2001, p 1792). Through such a dialectical view, Dikeç suggests that the *spatiality of injustice* 'implies that justice has a spatial dimension to it, and therefore, that a spatial perspective might be used to discern injustice *in space*', whereas the *injustice of spatiality* 'implies existing structures in their capacities to produce and reproduce injustice *through space*' (2001, pp 1792–1793, emphasis in original). Here the emphasis of spatial justice is not 'on space per se, but on the *processes* that produce space, and, at the same time, the implications of these produced spaces on the dynamic processes of social, economic, and political relations' (2001, p 1793, emphasis added). Dikeç's intention was to provide a framework for evaluating how injustices not only take place in space but how the formation of space itself affected social relations in unjust ways.

Such a dialectical interpretation of spatial justice brought a theoretical advance from the earlier notions of territorial justice analysed by geographers three decades earlier. Examining injustice dialectically allows one to frame relations of injustice by attending to the specific processes producing space and how these processes create 'dominant and oppressive permanences' in space and time which may reveal injustices. Although it stops short of a normative theory itself, Dikeç suggests that analysing the spatiality of injustice can help to illustrate the scale at which certain injustices prevail, but would require a normative theory which asserts how social relations and space ought to be differently produced in more just ways, such as with the right to the city (RTTC) thesis, to be discussed later.

Along the same lines of argument as Dikeç, Soja brought his dialectical thinking to bear on the concept of justice. Primarily as a response to prior geographical insights into justice theory, Soja's (2010) *Seeking Spatial Justice* asserts the necessity of space in theorising injustice. Justice has a 'consequential geography, a spatial expression that is more than just a background reflection or set of physical attributes to be descriptively mapped', Soja argues. 'The geography, or "spatiality," of justice ... is an integral and formative component of justice itself' (2010, p 1). More so than Dikeç, Soja asserts that space is to some extent constitutive of injustice. Returning to the socio-spatial dialectic, Soja (2010, p 5) suggests that 'the spatiality of (in)justice ... affects society and social life just as much as social processes shape the spatiality or specific geography of (in)justice'.

Throughout *Seeking*, Soja identifies examples which support his concern that not only do social relations produce unjust geographies, but geographies or 'spatialities' constrict human action; he draws on the examples of South

African apartheid, Palestinian occupation, gerrymandering, segregation and instances of environmental injustices to bolster his arguments. Soja's examples describe how social relations produce injustice in space and how space may inhibit certain rights and freedoms. This is not the same thing, however, as arguing that spatiality *itself* produces injustice as Soja does.[2] Rather, Soja's theory of a 'spatiality of (in)justice' promotes an ontology of space that has not been fully articulated. Soja too ultimately defers to the RTTC argument to illustrate his normative commitment to just spatial relations.

A supplement to Soja's ontological emphasis on spatiality is found in urban planning scholar Peter Marcuse's interpretation of spatial justice. Marcuse (2009) shares in the notion that space and social relations are intertwined, but does not concede that space actively produces injustice itself. Instead, and more concretely, Marcuse (2009, p 3) suggests five propositions which help identify spatial injustice. The first is that spatial injustice has two cardinal forms: 'the involuntary confinement of any group to a limited space – segregation, ghettoization – the unfreedom argument' and 'the allocation of resources unequally over space – the unfair resources argument'. The second is that spatial injustice is 'derivative of broader social injustice'. The third, that 'social injustices always have a spatial aspect, and social injustices cannot be addressed without also addressing their spatial aspect'. The fourth, that 'spatial remedies are necessary but not sufficient to remedy spatial injustice – let alone social injustices'. And finally, fifth, spatial injustice, because it is relative to social injustice, 'is dependent on changing, social, political, and economic conditions'. While Marcuse largely follows the socio-spatial dialectic, therefore, he guards against the idea that spatial injustices are fixed through spatial remedies alone, suggesting that a focus on spatiality itself cannot correct for injustices which derive from histories of social, political and economic injustice.

Marcuse's reservations notwithstanding, scholars continue to develop a spatiality of justice beyond Soja's socio-spatial dialectic. Post-structural legal scholar Andreas Philippopoulos-Mihalopoulos, for instance, laments the notion of spatial justice being merely 'a geographically-informed version of social justice' (2010, p 201). The issue with spatial justice theories, he argues, is that they are too 'aspatial' (2015, p 3). By attending to the 'spatiality' of space, therefore, he suggests that the concept has the potential to 'redefine, not only law and geography, but more importantly, the conceptual foundations of law and space' (Philippopoulos-Mihalopoulos, 2015, p 3). As such, he seeks to assert the essence of space to law and legal thinking as a heuristic, by defining spatial justice as 'the conflict between bodies that are moved by a desire to occupy the same space at the same time' (Philippopoulos-Mihalopoulos, 2015, p 3). Law and space cannot be separated from one another, he notes, but are conditioned together with one another. This 'non-dialectical' but rather 'interfolding' of law and space is what he terms the 'lawscape', defined as the way the 'ontological tautology between law and space unfolds as difference' (Philippopoulos-Mihalopoulos, 2015, p 4). It is through the lawscape that spatial justice 'emerges', not as an enforcement of law against bodies (broadly conceived) but of ever continuous negotiations of fixed positions in space. While Philippopoulos-Mihalopoulos acknowledges that spatial justice is a legal concept in some sense, it is important for him that spatial justice be seen as an assemblage of conflicts within and throughout space that do not have easily identifiable resolutions. More central to his approach to spatial justice is that it remains merely a question, a way of asking: what happens when bodies try to occupy the same space?

Such an understanding provides less a conception of spatial justice than a means of underscoring the distinction between justice and law. For Philippopoulos-Mihalopoulos, space is ontologically ordered, a continuum within which bodies 'rupture', and thus cannot hold, fixed (spatial) positions. Such an ontological conception of space, a notion of space as (un)structured, makes the search for a conception of spatial justice harder to identify. Instead, the idea of lawscape reiterates the notion that bodies share space with other bodies/non-human species. Not coincidentally, these were the insights at the core of Harvey's (2009) and especially Smith's (1994) work within geography: the idea that through our mere existence, we require some place, a space with which we must exist. The risk therefore in further reconceptualising ontologies of space is that the scale or even places of injustice become more difficult to identify.

A third approach to spatial justice seeks to more concretely identify spatial justice by developing a means of measuring it. Although they do not use the term itself, Israel and Frenkel (2018) advance a conceptual framework through which to assess the justness of spatial phenomena. Their framework seeks to evaluate injustices based on the notion of equality of life chances and opportunities, an idea central to the Capabilities Approach (see Chapter 8). To assess spatial injustice, they suggest we examine how 'socio-spatial structures and personal characteristics (i.e. living environment, habitus, and [cultural] capital forms) … may impair equality of capabilities' (Israel and Frenkel, 2018, p 659). For them, evaluating justice spatially is to assess the scales at which freedom of opportunities are structured as a result of the social relations which produce space. Capabilities thus are 'determined by a person's relative position in social space and a particular living environment (the *what* is). Or in other words, the quality and quantity of capital forms available to him or her in a given time and space' (Israel and Frenkel, 2020, p 4). But so too, they argue, are capabilities affected by the conditions of one's living environment. They stress it is necessary to examine how socio-spatial relationships impair or improve one's functionings within economic, political or social relations. Thus, the authors note that it is not enough to develop only a metric for evaluating injustices in space, or to evaluate the capacity to flourish; a normative commitment to assessing injustice must also address how those conditions were created and upheld.

By adopting the definition of well-being as the meaning of justice, Israel and Frenkel's spatial theory of well-being does not show us how analysing spatial relations helps us think about the conception of justice explicitly. Rather, it identifies spatial aspects of well-being which then must be connected to a theory of justice. Nonetheless, when social justice is equated with well-being itself, more specifically the capacity to flourish, such a spatial theory of injustice becomes narrower in definition.

As we have seen throughout the historical development of the concept, the diversity of approaches taken towards a notion of spatial justice has been vast. Before the term itself was in use, geographic thinking was applied to mainstream justice theories helping to identify critiques in distributive justice theorising. But so too did advances in spatial thinking change how space mattered for theorising justice. Herein, spatial justice can be seen as an analytical guideline for assessing how injustices are produced and maintained in the spatial processes constituting places. However, as a collective body of research, it comprises less a normative theory of justice in itself and more so a means of framing the site and scale of injustices. And in doing so, the analytical utility of a spatial approach to justice is that it provides conceptual tools useful in making evaluations about situations of justice or injustice. In this way, spatial justice may be complemented by other geographic theories which rely upon more normative theorising.

Connecting the concept of justice with theories of space

We saw in the previous section how the concept of spatial justice has been developed as an analytical framework. As a means of framing analyses, spatial justice approaches often lack more normative means of articulating how unjust relations should be made more just. That is, spatial justice approaches often do not articulate how some unjust process ought to be made more just as much as they explain how certain circumstances or situations create unjust conditions (Barnett, 2011). Thus, to further the connection between spatial thinking and normative theories related to justice, this section details key theories using spatial concepts in order to illustrate the connection to spatial theories of justice more broadly.

Perhaps the most prominent spatial and scalar theorising related to, but not explicitly drawing from, justice theories is scholarship on RTTC. The RTTC concept has been widely developed within and outside of the discipline of geography. A set of ideas put forth by Marxian philosopher Henri Lefebvre in the socio-political upheaval of 1968, RTTC is simultaneously a critique of the urbanisation processes producing the modern capitalist city while also an assertion of the moral right to democratise the city for those excluded from those processes. The problem with the capitalist city for Lefebvre (1996) was how the capitalist mode of production was eliminating the opportunities for the working class to shape the development of the city broadly conceived. He saw the hegemony of exchange value over use value within the neo-capitalist urbanisation process as a dominating force dispossessing the working class from 'the urban' (Lefebvre, 1996, p 179). His call for a 'right' to the city, or to urban life more generally, therefore, was the right of all to inhabit and appropriate the city. 'The right to the city manifests itself as a superior form of rights: right to freedom, to individualization in socialization, to habitat and to inhabit. The right to *oeuvre*, to participation and *appropriation* (clearly distinct from the right to private property), is implied in the right to the city' (Lefebvre, 1996, pp 173–174, emphasis in original). That is, all urban inhabitants have not only the political, but moral right to occupy and actively shape the spaces of the city to satisfy their collective desires and not just the desires of capital.

Perhaps reflecting the influence of Marx on his thinking, Lefebvre does not directly connect RTTC with the concept of justice. Indeed, much RTTC scholarship implies there is injustice without rationalising what justice means. For instance, RTTC implies a push for a more democratic social order generally (Purcell, 2008) with its collective goal aimed at 'democratic control over the production of and utilization of the surplus' (Harvey, 2008, p 37). Hundreds of articles drawing from RTTC as a concept examine, for instance, issues of gender discrimination in urban space, housing segregation, quality and access to sanitary infrastructures, education, citizenship, and much more. Though generally not rooted in the concepts of normative justice theories, the implication of this voluminous body of scholarship is that by democratising the social, economic and political processes of urbanisation, justice will be advanced.

Some spatial thinkers have made connections with RTTC with social justice. Don Mitchell's (2003) work on homelessness and public space, for instance, argues that the struggle to maintain the presence of public spaces, spaces of representation that enable difference to flourish as a necessary component of urban life, is fundamental to a socially just city. For him, public space *is* the space of justice. 'It is not only where the right to the city is struggled over; it is where it is implemented and represented' (Mitchell, 2003, p 235). And in this sense, public space represents 'a gauge of the *regimes* of justice extant at any particular moment' (Mitchell, 2003, p 235, emphasis in original). For Mitchell, to examine the ways in which space is produced and maintained in the city is to begin to address how injustice is structured and reproduced within the world. However, there remains much room for RTTC scholarship to draw from justice concepts in order to better rationalise how RTTC's critique of modern capitalism is about the injustice of space and its production.

A second theory attending to how space and scale help identify situations of injustice is addressed in Marxian theories critiquing accumulation by dispossession (ABD). ABD is a theory concerned with how space becomes integral to the circulation of capital throughout the landscape. In general, accumulation is the process through which capital is reproduced on an expanding scale by reinvesting surplus value from previous rounds of production. For Marx, 'primitive' accumulation, through the enclosure of the commons, was the precondition for the capitalist mode of production. Marx saw how primitive accumulation 'entailed taking land ... enclosing it, and expelling a resident population to create a landless proletariat, and then releasing the land into the privatised mainstream of capital accumulation' (Harvey, 2003, p 149). But Harvey argues that Marx understood these actions as relegated to the past, not as ongoing processes temporally and spatially. Thus, Harvey argues that in the contemporary neoliberal moment, it is the continued dispossession through privatisation of commonly owned goods that is at issue (2003, pp 137–182). Through ABD, therefore, Harvey develops a certain spatiality to explain Marx's thinking, by showing how capital accumulation requires that geographical space be made and remade for its extension, and in so doing, produces socially and ecologically unequal relations throughout space.

Directly related to ABD is the notion of 'uneven development'. Uneven development is a process by which development varies in time and space based on the capitalist mode of production. Uneven development connotes that capital develops unevenly because it seeks out spaces that better maximise its ability to produce profit. Uneven development can be seen at a variety of scales; from Global North to Global South; from regions within a nation state; or from neighbourhood to neighbourhood within a given city. Marxist geographers like Neil Smith (1996, 2008) have argued that the capitalist mode of production depends on the spatial-temporal unevenness in the production of space. Unevenness is a necessary outcome of capitalist production because capital requires a site of investment within which to engage in production, while at the same time, capital must remain mobile to circulate as value and thus remain available for investment elsewhere with higher rates of profit.

Although the concepts of ABD and uneven development do not explicitly engage with a conception of justice, they relate to justice in the same sense that justice is implied within Marxist analysis of the capitalist mode of production (see Chapter 5). That is, injustice for Marx relates to how labour is alienated from its own means of production. Nonetheless, these concepts have been critical for explaining how social inequities arise spatially by illustrating the ways in which capital produces space for its own reproduction.

The concepts can be seen as explanatory theories helpful in situating the conditions of injustice within a broader theory of justice.

Spatial justice and the social sciences

Theories of spatial justice are perhaps most useful for social science researchers for how they identify a *domain* of justice (see Introduction). That is, a spatial justice framework examines how injustices arise within or through spaces by helping define the scales at which injustices take place. For example, a spatial justice lens may help to frame the ways in which the grazing patterns of reindeer in Scandinavia conflict with private use-rights of property, given how the needs of reindeer enact a spatiality illegible to private property requirements (Brown et al, 2019). In turn, the spatial production and enactment of private property can affect the livelihoods of Indigenous peoples who remain dependent upon herding reindeer. So too can the concept of scale be usefully applied to assess issues of deforestation. Large-scale deforestation is multi-scalar in consequence. Deforestation reduces the ability of forests to function as carbon sinks necessary for ameliorating greenhouse gas increases which affect atmospheric change, changes felt throughout the globe. But so too may an attention to scale illustrate how political-economic decisions at one scale affect those of another. To stay with the forestry example, the extent of property rights of the private forestry industry may affect the ability of local populations to shape public discourse over how best to pursue timber production in more sustainable ways.

These are but two examples. A spatial justice framework allows for researchers working on a variety of social and ecological issues a useful analytical tool for identifying conditions where injustices take place. This also means that spatial justice is not a means of defining whether something is just or unjust. For, it does not provide a normative assessment of why something is considered just or unjust. Instead, the utility of a spatial justice approach lies in the way it analyses potentially unjust conditions and how they are *emplaced*. With an acute attention to scale, a spatial justice framework can help to identify how a given situation of injustice takes shape in a given place as well as how that injustice is affecting others beyond the immediate location of the injustice(s). In applying such a framework, therefore, researchers may draw analytical insights about the forms in which injustice is produced and maintained in order to then advance more normative assertions about why something is injustice and what can be done to correct for it.

Notes

[1] See for example the pivotal paper by Marston et al (2005) which calls for the eradication of scale in human geography, and the many subsequent responses to that assertion, such as Leitner et al (2007) and Jonas (2006).

[2] Soja's notion of the socio-spatial dialectic has received much criticism since it was first proposed in 1980. Some have been critical of the notion that spatiality can *affect* social relations without veering into a certain type of determinism. Geography has an ominous legacy with environmental determinist thought, which sought to justify racial inferiority with one's environmental circumstances. Soja is well aware of this legacy and states that the socio-spatial dialectic does not determine social attributes or behaviors. Thus, the charge against Soja is more so rooted in spatial fetishism and a reductionism of the totality of social space into separate registers or 'parallel structures for whose investigation one could or should legitimately claim a clear epistemological autonomy' (Lopes de Souza, 2011, p 76).

References

Barnett, C. (2011) Geography and ethics: Justice unbound. *Progress in Human Geography*, 35(2), 246–255.

Brown, K., Flemsæter, F. and Rønningen, K. (2019) More-than-human geographies of property: Moving towards spatial justice with response-ability. *Geoforum*, 99, 54–62.

Dikeç, M. (2001) Justice and the spatial imagination. *Environment and Planning A: Economy and Space*, 33(10), 1785–1805.

Fainstein, S. (2010) *The Just City*. Ithaca, NY: Cornell University Press.

Gregory, D. (2009) Space. In D. Gregory, R. Johnston, G. Pratt, M. Watts and S. Whatmore (eds), *The Dictionary of Human Geography*. Malden, MA: Blackwell. p 708.

Harvey, D. (1992) Social justice, postmodernism, and the city. *International Journal of Urban and Regional Research*, 16(4), 558–601.

Harvey, D. (1996) *Justice, Nature and the Geography of Difference*. Malden, MA: Blackwell Publishers.

Harvey, D. (2003) *The New Imperialism*. Oxford, UK: Oxford University Press.

Harvey, D. (2008) Right to the city. *New Left Review*, 53, 23–40.

Harvey, D. (2009) *Social Justice and the City*. Athens, GA: University of Georgia Press.

Israel, E. and Frenkel, A. (2018) Social justice and spatial inequality: Toward a conceptual framework. *Progress in Human Geography*, 42(5), 647–665.

Israel, E. and Frenkel, A. (2020) Justice and inequality in space: A socio-normative analysis. *Geoforum*, 110, 1–13.

Jonas, A.E.G. (2006) Pro scale: Further reflections on the 'scale debate' in human geography. *Transactions of the Institute of British Geographers*, 31(3), 399–406.

Lefebvre, H. (1996) The right to the city. In E. Kofman and E. Lebas (eds), *Writings on Cities*. Malden, MA: Blackwell Publishing, pp 63–181.

Leitner, H. and Miller, B. (2007) Scale and the limitations of ontological debate: A commentary on Marston, Jones and Woodward. *Transactions of the Institute of British Geographers*, 32(1), 116–125.

Lopes de Souza, M. (2011) The words and the things. *City*, 15(1), 73–77.

Marcuse, P. (2009) Spatial justice: Derivative but causal of social injustice. *Spatial Justice*, September. https://www.jssj.org/article/la-justice-spatiale-a-la-fois-resultante-et-cause-de-linjustice-sociale/

Marston, S., Jones III, J.P. and Woodward, K. (2005) Human geography without scale. *Transactions of the Institute of British Geographers*, 30(4), 416–432.

Mitchell, D. (2003) *The Right to the City: Social Justice and the Fight for Public Space*. New York: Guilford Press.

Moroni, S. (2020) The just city. Three background issues: Institutional justice and spatial justice, social justice and distributive justice, concept of justice and conceptions of justice. *Planning Theory*, 19(3), 251–267.

Philippopoulos-Mihalopoulos, A. (2010) Spatial justice: Law and the geography of withdrawal. *International Journal of Law in Context*, 6(3), 201–216.

Philippopoulos-Mihalopoulos, A. (2015) *Spatial Justice: Body, Lawscape, Atmosphere*. New York: Routledge.

Pirie, G.H. (1983) On spatial justice. *Environment and Planning A: Economy and Space*, 15(4), 465–473.

Przybylinski, S. (2022) Where is justice in geography? A review of justice theorizing in the discipline. *Geography Compass*, 16(3), 1-12.

Purcell, M. (2008) *Recapturing Democracy: Neoliberalization and the Struggle for Alternative Urban Futures*. New York: Routledge.

Smith, D.M. (1994) *Geography and Social Justice*. Oxford, UK: Blackwell.

Smith, D.M. (2000) Social justice revisited. *Environment and Planning A: Economy and Space*, 32(7), 1149–1162.

Smith, N. (1995) Social justice and the new American urbanism: The revanchist city. In A. Merrifield and E. Swyngedouw (eds), *The Urbanization of Injustice*. London: Lawrence and Wishart, pp 117–136.

Smith, N. (1996) *The New Urban Frontier: Gentrification and the Revanchist City*. New York: Routledge.

Smith, N. (2008) *Uneven Development: Nature, Capital and the Production of Space*, 3rd edn. Athens, GA: University of Georgia Press.

Soja, E.W. (1980) The socio-spatial dialectic. *Annals of the Association of American Geographers*, 70(2), 207–225.

Soja, E.W. (1989) *Postmodern Geographies: The Reassertion of Space in Critical Social Theory*. New York: Verso.

Soja, E.W. (2010) *Seeking Spatial Justice*. Minneapolis, MN: University of Minnesota Press.

Young, I.M. (1990) *Justice and the Politics of Difference*. Princeton, NJ: Princeton University Press.

13

Landscape Justice

Don Mitchell

Introduction

While the concept of landscape justice is relatively new, the practice of it may not be. As Kenneth Olwig (1996) has shown, there has long been a tight link between landscape (as a certain kind of place) and systems of justice, especially, though hardly exclusively, in the Nordic countries. To understand this link, as well as to understand the range of contemporary approaches to landscape justice, however, we will first consider the complex meanings of *landscape*, because it is hardly the self-evident word it often seems to be, and its very complexity as a term *and* as a phenomenon has shaped the way in which justice is theorised and practised in relation to it.

Landscapes and justice: key ideas

In everyday usage, 'landscape' is typically understood to be a 'stretch of inland scenery' as standard dictionary definitions have it. In this sense, landscape is both the view, which licenses landscape painting and photography, and what is viewed, which licenses its metaphorical use, as in 'the political landscape', or 'the intellectual landscape', but which more specifically indicates an area, territory, space or morphology (Sauer, 1925). Landscape is both the representation and the represented, which, as we will see, is crucial for discourses of landscape justice (Mels, 2016). Landscape differs from 'environment', again with crucial implications for justice theorising, in that the latter is more 'objective' (indicating either the 'surrounds' of our lives or the ecological substrate of a place) while the former is both humanly produced and more 'subjective' in the sense that landscape is always imbued with meaning. In the words of the European Landscape Convention (ELC, discussed more fully in what follows), the landscape is 'an area, as perceived by people, whose character is the result of the action and interaction of natural and/or human factors' (ELC, ch 1, art 1; Déjeant-Pons, 2006), a definition that accords well with its historical usage within geography, landscape architecture and other spatial disciplines (Sauer, 1925; Meinig, 1979; Wall and Waterman, 2019), as well as more popular explorations into vernacular landscapes (Jackson, 1984). Landscape in this definition is the humanly transformed environment, imbued with meaning, that serves as the 'infrastructure' for everyday life (Nye, 2010).

In his landmark theoretical statement, 'The morphology of landscape', Carl Sauer (1925) declared that scientifically, landscape is a 'naïvely given section of reality'. Yet since landscape is 'representation' as much as it is the 'represented', Sauer's argument cannot hold. Beginning in the 1980s, a significant line of research arose examining the specific ways within which landscape *is* representation and entails a politics of representation. As Denis Cosgrove (1984; 1985) and others showed, the 'landscape way of seeing' (cf Berger, 1972), closely linked to the invention of single-point perspective and various cartographic technologies, was a Renaissance innovation, tied to the demarcating of landed estates in Venice's *terra firma* and in the western European Low Countries. Transforming land into landscape, as a well-ordered view, as *scenery* (Olwig, 2002; 2019), simultaneously entailed transforming land into *property* and was thus closely linked to early capitalist enclosures of the commons, the dispossession of peasants from customary lands, and the dissolution of the monasteries (Fields, 2017). This line of research has shown how, historically, the ability to depict the land as landscape (and property) became the ability to *remake* the land as landscape, not only through the invention of landscape design (which was crucial) but also through the legal and violent ability to dispossess. It also became tightly linked to politics and a set of practices of representation that knitted landscape depictions (and forms) to expressions of (particularly national) identity (Daniels, 1993; Matless, 1998). If 'landscape' signifies a kind of belonging, as a good deal of contemporary phenomenological landscape theory argues (for a review: Wylie, 2007), then it does so precisely because it is exclusionary. Landscape in this sense incorporates alienation right into its essence. To see landscape as a naïvely given section of reality, or *innocently* as 'an area, as perceived by people, whose character is the result of action or interaction of natural and/or human factors' is simply impossible, and threatens to perpetuate rather than challenge injustice.

Yet the turn to representation, as Tom Mels (2016) explains, came at some cost to the understanding of landscape as a structured, built form. Mitchell (1996; 2003a; 2003b; 2008; 2012), therefore, sought to rehabilitate and reorient Sauer's interest in *morphology* through the development of historical-materialist analyses of the relations of labour that go into landscape's making, while also remaining attentive to its representational aspects. In this view, landscape is *built environment* (Harvey, 1982) that both internalises the relations of labour that produce it *and* significantly determines the conditions of possibility for future labour practices. Rooting his arguments in Marx's analysis of capitalism, Mitchell argued that landscape is necessarily fetishistic and alienating (even as it is also exactly, as the ELC defines it, an 'area, as perceived by people' which is the 'result of the action and interactions' of human and natural processes). Reduced to slogans, Mitchell's argument is that (1) landscape is 'dead labour' (the labour of its making, internalised and concretised); (2) landscape is deceptive (it hides the relations of its making); and (3) landscape *is* power (it results from, and has powerful influence over, the struggles that go into its making). In this view, landscape encapsulates the 'actually-existing' state of justice, as expressed through relations of production, and thus sets *morphological* limits to how struggles for greater justice may unfold.

Olwig (1993; 1996; 2002; 2019) finds such an argument incomplete. For, his research has shown, there is another, crucial history to landscape. In pre-capitalist Nordic countries, in particular, landscape (*landskap*, *landskab*) was historically a *territory*, often operating under different customary laws than surrounding feudal territories. A landscape was, in Olwig's (1996, p 311) words, 'an area carved out by ax and plough which belongs to the people who have carved it'. Landscapes were political spaces in the sense that they were both a territory defined by a people and a territory that shaped a polity. Landscapes were (proto-)democratic spaces, typically centred on a *ting* (parliamentary or ceremonial space, often a circle of rocks) and thus a space of representation, in the sense that it was within and as part of a landscape that one could represent one's interests (or have them represented). In this sense, landscapes were spaces of justice. 'Landscape justice' was thus lived in place and enacted through custom. Olwig does not deny the importance of usurpation, enclosure, the rise and hegemony of capitalist property relations, alienation, or the growing importance of pictorial definitions of landscape (and their associated relations of power) in modern, capitalist history, but rather insists that this more 'substantive' meaning of landscape, and its association with practices of justice, continued to persist within, and contest, these more dominant meanings and practices, thereby significantly shaping them.

Olwig's work in the 1990s was vital for injecting explicit consideration of justice into landscape studies. The need for such consideration was amplified by George Henderson (2003) in his reframing of the heritage of J.B. Jackson's concern with the vernacular landscape as a question of how to conceive of more just landscapes (about which more in the next section). Though not always discussed in the language of, or in explicit relation to, theories of justice, Mitchell's work on labour and Richard Schein's (1997; 2006) work on race focused on landscape's injustices, which helped turn landscape analyses in more normative directions. Gunhild Setten (2004) widened the normative perspective by focusing on what she called moral landscapes. The writing, signing and evaluation of the ELC in the same years focused attention on questions of participation and scale, while raising concerns over localism and nationalism (points developed in the section on the ELC later). Shelly Egoz's (Egoz et al, 2011; 2018) calls for a *right to landscape* and *landscape democracy* sought to suggest means by which these matters of participation and scale could be turned in the direction of more just and inclusive landscapes (see also Jones, 2009; 2016).

Yet outside debates over participation and its limits, there have been surprisingly few intellectual or even political debates over the content of landscape justice. This is likely because that content has yet to be theorised in any rigorous fashion. Though Mels (2016; Mels and Mitchell, 2013), in particular, has sought to synthesise landscape theories with theories of oppression, domination and exploitation emanating from the work of Iris Marion Young (see Chapter 4) and the triumvirate of distribution, recognition and representation developed by Nancy Fraser in her debates with Axel Honneth, his arguments have yet to be taken up and further developed. At best, 'landscape justice' is aspirational and embryonic.

Landscape justice: key debates and critiques

Nonetheless, certain key tenets of landscape justice theorising can be discerned.

Landscape, everyday life and justice

In an essay reflecting on the importance of J.B. Jackson's work on vernacular landscapes for landscape studies more generally, Henderson wrote the following:

> Different landscape concepts rest on different ontologies, on varying notions of what the world is like and what's worth pointing out about it. ... And very worthwhile for new conceptions and studies of landscape will be a discourse that defines landscape as a necessary and integral component of more just social relations. What is also needed is a concept of landscape that helps point the way to those interventions that can bring about much greater social justice. And what landscape study needs even more is a concept of landscape that will assist the development of the very idea of social justice. To achieve this, geographers and other landscape analysts will need to engage in a more sustained conversation with the disciplines

of moral and political philosophy concerning the enumeration of basic human rights and the modes of their defense. (Henderson, 2003, p 196)

He goes on to argue that studies of landscape must find ways to address:

[T]he concern for security, safety, and joy in one's work; the struggle for wages that guarantee a share in the good life; the question of who gets to decide what work is, what work gets done, and what goods get made; the fight against excessive personal and corporate accumulation of wealth and power; the idolatry of the market. (Henderson, 2003, p 196)

And he concludes by saying:

The list could go on, but the study of landscape, that thing which so often evokes the plane on which normal, everyday life is lived – precisely *because* of the premium it places on the everyday – must stand up to the facts of a world in crisis, to the fact that the condition of everyday life is, for many people, the interruption of everyday life. (Henderson, 2003, p 196, emphasis in original)

Part of that good and everyday life is the right to inhabit a beautiful landscape, but, citing J.B. Jackson, Henderson (2003, p 197) made it clear that 'any definition of the beautiful landscape would have to include the full participation of all and the economic means to do so'.

Landscape injustice

In a 'progress report' on cultural geography in *Progress in Human Geography*, Mitchell (2003a) picked up on Henderson's argument and called for a full synthesis of landscape study with the theorisation and struggle for social justice, though without responding to Henderson's call for a direct engagement with moral and political philosophy. In addition to pointing to the important work of a group of Nordic scholars (discussed later on), Mitchell sought to lay the groundwork for a *political-economic* and *historical-materialist* approach to assessing the social, historical and cultural processes that led to injustice in the landscape. In particular, he argued that relations of labour were decisive in the possibility for landscape justice (Mitchell, 2003a). He followed this up a few years later by laying out a set of 'axioms', or precepts, for developing a theory of landscape suitable for understanding the political-economic relations of social justice – or more accurately injustice (Mitchell, 2008). Mitchell argued that any study of landscape that might be able to point the way towards 'greater social justice' (in Henderson's words), had to be based on these foundational axioms:

- Landscapes are produced, actively made (not merely what Peirce Lewis [1979] called 'unwitting autobiographies'); they are physical interventions in the world.
- Landscapes are functional; that is, they play a mediating role in relations of production and social reproduction.
- Landscapes are not only local (as in the ELC definition), but are the result of processes operating at a myriad of geographical scales.
- Landscapes are *historical*; the production of landscape, as well as its evolving use and meaning, is a historical (as well as a geographical) process and thus the study of landscape must be oriented towards understanding that history.
- Landscape is power; it does not only mediate, but also shapes social relations even as it is a product of the power relations at the heart of any social process.
- 'Landscape is the spatial form that social justice takes' (Mitchell, 2008, p 45), which is to say that, given the previous points, it encapsulates the *actually-existing* relations of justice; to the degree this is true, any effort to create a more just world will require a significant transformative remaking of the geographical landscape, not 'merely' the remaking of social relations and institutions.

Whatever the value of these axioms for understanding the spatial determinants and forms of social justice and injustice, in this and other work (with only slight exceptions: Mitchell, 2003b; Mels and Mitchell, 2013), Mitchell never really engaged with the sort of moral and political philosophy Henderson pointed to. Instead, and in common with most geographers, he simply took 'social justice' as an unexamined 'good', a self-evidently desirable normative outcome that one will know when one sees it. If there is a theory of justice behind his efforts to construct a theory of landscape rooted in social justice, it is probably a basic sense of 'justice as fairness' (Rawls, 1971), though cut through with a kind of historical-materialist scepticism towards ever achieving the conditions of possibility for such fairness. (Shorn of this scepticism, the sense of landscape justice as justice-as-fairness is probably the dominant way of understanding the matter; see for example, Dalglish et al [2018].) Put another way, much of Mitchell's work has been less concerned with landscape in relation to justice (or 'landscape justice') than it is with understanding how landscapes continuously instantiate injustice, unfairness and structural violence (Mitchell, 2008; 2012), and what sort of struggles might be necessary to combat this.

In this Mitchell has not been alone. The focus on *injustice*, rather than on theorising justice, is predominant within landscape studies, perhaps for good reason (cf Barnett, 2017), and certainly not without some significant conceptual developments that allow for more precise understandings of the historical-geographical processes that instantiate injustice as a material fact in the landscape. For example, Miguel Torres Garcia and his collaborators (2020, p 618; citing Setten and Brown, 2013) have usefully distinguished between *landscapes of injustice*, which are 'the outcome of, and may reveal, inequalities', and *landscape injustice* which 'cloaks [inequalities] under an exclusive discourse which naturalizes them and makes them seem inevitable, which leads to injustice'. The implication here is that social scientists interested in understanding the preconditions for social justice need to understand how landscapes are always *both* landscapes of injustice and instantiations of landscape injustice.

Landscape as a place of justice

But social scientists also have to understand how landscapes are, or can be, places of justice. The group of Nordic scholars Mitchell pointed to in his 2003 'progress report' have long focused on just this question. For Olwig in particular, this question has entailed developing an understanding of how landscapes are produced, practised and transformed through the interrelation of customary and statutory law, which is to say, the substance of landscape, like the substance of justice, is legally shaped. Olwig's extensive corpus is at once geographically wide-ranging (examining Nordic, Caribbean, North American, British, Greek and other landscapes) and singularly focused on the philological and historical excavation of landscape *meanings* as they have developed and shifted over time and recursively shaped social life (Olwig, 2019). The question of justice is central to this work because (as noted) the substance of landscape itself is, in the Nordic countries, 'justice', but in a very specific way.

Olwig grounds his arguments on the distinction Aristotle (1934) made between 'two kinds of justice': (1) 'natural justice', which is universal 'and has the same force everywhere'; and (2) 'conventional justice', which arises 'in particular places and times and tends to grow in force' (Olwig, 2019, p 48n22). Olwig argues that each type of political justice 'engenders its own political landscape', which in his usage is *not* metaphoric. Rather, one political landscape is 'de-centered in universal space' (his example is the township and range demarcation of trans-Mississippi western America), while the other, through customary usage of the land and sedimented legal custom is 'centered on the particularity of place' (his example is the New England village with its town commons and meeting hall). These two kinds of justice, and the associated landscapes they give rise to, are always in dialectical tension, contradictory but also potentially complementary. Natural justice and conventional justice have spatial forms and the dialectical dance between them gives rise to actually-existing landscapes. And yet, beyond this *political* sense of justice (the application of law), Olwig has relatively little to say about the substance of justice. For him, justice is always in a state of becoming, defined by the tussle between space and place, custom and natural law, the local and the universal.

The moral landscape and productive justice

Another Nordic geographer, Gunhild Setten (2004; 2020; Setten and Brown, 2009; 2013), agrees that justice is always in a state of becoming, but has placed landscape practice in a more central position than Olwig by focusing on the ways in which landscapes are always *moral* landscapes. She has shown how landscape is always infused with competing moralities, competing convictions of what is good and bad. 'Shared moral assumptions' within groups concerning right and wrong – what Setten labels a 'moral order' – shape behaviour in the landscape through 'contested codes of conduct' and divergent 'ordering practices'. Landscapes themselves 'are the product of rules aimed at ordering and producing practices that ultimately are cast as natural or unnatural, moral or immoral'. 'Landscape narratives' thus become key ingredients in 'justifying different types of conduct' and land uses (Setten and Brown, 2009, p 113).

In this view, 'justice' is internally related to a normative order, not something that stands outside and defines it. There is, thus, a certain relativity in moral – or justice – claims, while they are also at the same time grounded within specific ways of knowing and historically developed practices. The moral orders governing farmers' or hunters' use of the land might differ in significant ways from, for example, those of environmental bureaucrats and planning officials (Setten, 2004). For Setten and Brown, the crucial point is

> that people try to *do* the landscape in different ways; that there are different judgments about the appropriateness of the doing and that the landscape is both implicated in the doing and the passing of judgements, and hence there are implications for who is included and excluded, and in what sense. (Setten and Brown, 2013, p 244, emphasis in original)

Drawing on geographer Richard Schein (2006), whose work has focused on the relationship between race, racism and the landscape, Setten and Brown (2013, p 243) argue that 'it is *always* possible to think about landscape and social justice, *even* in one's everyday environments' (emphasis in original). Important for Setten's arguments about the moral landscape, however, is a distinction she makes (citing the Marxist political ecologist James O'Connor [1998] though giving the argument an important feminist twist) between *distributive* justice (defined as 'mechanisms of distribution and their fairness') and *productive* justice. To some degree, productive justice might be understood as a species of what Nancy Fraser has identified as the issue of (political) representation (see Chapter 4): a question of 'involvement [in] and control over choices and decisions' (Setten and Brown, 2013, p 244). But in relation to landscape, it is in fact a broader matter than Setten and Brown indicate: it is a question of control not only over choices and decisions, but especially over choices and decisions concerning *how the landscape shall be shaped* – how it will be *produced* – and to what effects.

The European Landscape Convention and the limits to participation

The question of 'productive justice' is evident, if in a somewhat different way, in the articles of the ELC, as are questions of distributive and procedural justice. Beginning in 1994, the Council of Europe undertook a process of developing and codifying a set of landscape precepts that, when transformed into a Convention, would oblige signatory parties (European states) to certain forms of action in relation to the landscape. The ELC was signed in 2000, came into force in 2004, and by 2020, 30 of 47 European countries had committed themselves to it. In relation both to the substance of the landscape and to procedures for developing, transforming or preserving it, the Convention encapsulates a set of relationships between landscape and justice, shaped by competing moral orders (to use Setten's terms). Having its origins in a felt need to conserve and protect valuable landscapes, the ELC was broadened during negotiations to include all landscapes, including degraded ones. The scope of the Convention was expanded to include landscape planning and landscape management as well as landscape conservation, and the broad definition of landscape being 'an area, perceived by people' that results from human–nature interactions was adopted (Bruun, 2016, p 11). Perhaps most significant, however, is that the ELC enshrines a particular vision of a *procedurally* just process for governing landscape production and transformation.

First, the ELC requires parties to 'recognize landscapes *in law* as an essential component of people's surroundings, an expression of the diversity of their shared cultural and national heritage, and a foundation for their identity' (quoted in Jones, 2009, p 233, emphasis added). Second, the ELC 'promotes' active participation, principally though not exclusively, in the form of 'consultation' (Jones, 2009, pp 234–235). And third, it 'confirms the principle of subsidiarity, whereby decisions should be taken at the lowest practical administrative level, thus enhancing local democracy' (Jones, 2016, p 119). The writing and adoption of the ELC spawned a deep flood of literature assessing its importance and shortcomings, debating its promise as a guiding text for state and political entanglement with the landscape, and examining how it has (and especially has not) affected landscape planning in light of its articles. Behind all this work is a general, if not always explicitly articulated, sense that if fully implemented, the ELC would lead at minimum to better landscape policies, but more expansively potentially to the production and protection of more just landscapes.

What that justice might be, however, is never really outlined, except insofar as the ELC's provisions for public participation have the potential to encourage a more procedurally just planning process. Michael Jones (a third member of the group of Nordic scholars mentioned earlier) is the one who has examined this potential most closely (but see also Olwig, 2009), subjecting the ELC and the 'guidance' that accompanies it to careful scrutiny. He shows that while the ELC rhetorically promotes broad participation, in practice, especially when the words of the Convention are compared to the words of its own 'Explanatory Report', public participation is reduced to predefined 'stakeholders', and even then always subordinated to the rule of experts (Jones, 2009). Consultation, rather than deep public engagement, is the order of the day. In this sense, the *potential* for a broadly procedurally just system of public determination of landscape productions and practices is cut through and reshaped by structures of power that tend to reinforce the status quo rather than lead to transformative outcomes.

Similarly problematic, as Jones and a number of others have pointed out, is that the ELC's notion that landscape is an area *as perceived by people* is usually defined as *local* people – imagined, locally rooted communities – frequently to the exclusion of immigrants, migrant workers and distant others who might have a 'stake' in it (or in fact be the very ones who produce and maintain it). There is a localism, even a tightly bound communalism, built into the structure of the Convention that is potentially troubling and even damaging to the claims (for involvement, access, livelihood, and so forth) of presumed 'strangers'. The ELC has the potential to reinforce a regressive form of the link between people and place that, for example, Olwig (1993; 2019) argued, was central to the rise of the Nordic notion of landscape: a kind of 'blood and soil' politics, the goal and outcomes of which are all too well-known in European history. If the mandate for participation is limited on the one hand by its subordination to the rule of experts, it is undermined on the other by a potentially limited scalar reach that makes landscape the exclusive province of those who (presumably, but rarely ever actually) have 'carved it out with ax and plough'.

The right to landscape

For these reasons, Shelley Egoz argues that the ELC in particular, but landscape politics more generally, needs to be understood in relation to what she calls *the right to landscape*. Beginning with a conference at Cambridge University in 2008 (Egoz et al, 2011) and extending through a long list of publications, Egoz has argued that the right to landscape needs to be counted among the fundamental human rights, or rather that the landscape needs to be understood as a primary foundation for, and expression of, the struggle for human rights. The right to landscape is located where landscape, as 'physical elements and resources' as well as 'social, cultural, and economic values' overlaps with human rights, defined as 'rights that support existence' as well as 'rights that support dignity' (Egoz et al, 2011, p 6). In this sense, the right to landscape names a *space of justice* which is defined by a right of access to necessary resources, social, cultural and economic goods, and a right to a meaningful life. Egoz's arguments concerning rights to landscape are also significantly scale-sensitive and throughout her writing, she is especially attentive to the needs, interests and roles of 'outsiders' (immigrants, migrants, visitors, the marginalised) in the landscape, and particularly the need to promote, from the outset, *their* right to the landscape (Egoz and De Nardi, 2017). For Egoz, the right to landscape is the starting point for 'landscape democracy', which itself is understood as the 'path to spatial justice' (Egoz et al, 2018). As they emerge over the course of the edited book from whose title they are drawn, 'landscape democracy' and 'spatial justice' align most closely with Nancy Fraser's (2008) conceptualisation of justice as consisting of fair distributions, true recognition and full representation.

Landscape, justice and the logics of representation

The evolution of landscape theory to this point, where questions of democracy and justice are at its heart, is something that Mels has sought to take account of. He is one of the few landscape geographers to answer Henderson's call to engage deeply with justice philosophy (see also Mason and Milbourne [2019], which engages normative theories of justice, but is more about 'energy justice' than 'landscape justice', despite its title). Mels has reviewed Rawlsian, Marxian and feminist theories of justice to show how they relate to landscape, focusing particularly on Iris Marion Young's theories of oppression and related arguments concerning structural violence to show how these help expose the unjustness of landscapes as well as the close tie between landscape justice theorising and the larger body of literature on environmental justice (Mels and Mitchell, 2013). But in a special issue for the journal *Landscape Research* which sought to establish an agenda for landscape research (Jorgensen, 2016), he has reformulated the history of theoretical development in geographical landscape research to show how different epistemological and ontological orientations have different 'logics of representation' at their core.

Representation, Mels (2016, p 417) argues, is 'a core concept of justice' which is always 'entwined with the social and material struggle over the right to landscape'. The 'new cultural geographers' (like Denis Cosgrove, discussed earlier) of the 1980s and 1990s, Mels (2016, p 417) contends, were concerned with *cultural representation* (cf Hall, 1997), wherein the politics of representation were understood to be crucial to 'the manoeuvres of discursive power, hegemonic ways of seeing, identity formation and modernity, etc.' and thus deconstructing these was vital for understanding how power was built into landscapes. By the turn of the 21st century, however, such concerns were being supplemented (or maybe even surpassed) by a concern – articulated, Mels (2016, p 418) suggests, in different ways in Olwig's and Mitchell's work – with *political representation*, or the 'right to be represented' (which, of course, also clearly echoes Egoz's arguments). In this view, 'landscape was not just a cultural representation but the material expression of the struggle over justice, polity and the peoples' cry and demand for a place of representation'. In turn, this concern with representation closely aligned with developments in feminist-socialist political philosophy as they were being worked out by Fraser (1997; 2008) and Young (2000).

Mels' concern, however, was less with landscape and landscape justice *as such*, and more with the potential that research into the representational logics of landscape offered for understanding environmental justice. In particular, Mels (2016, p 422) pointed to how landscape theory developed through the logic of *political representation* has much to offer environmental justice theory in terms of how to better understand:

- *Place contexts* and especially the importance of historical context.
- *Spatial scale* and especially the complex scales of justice (Fraser, 1997).
- *Political representation* itself, especially as it is entangled with both economic and cultural injustice.

Together, these three aspects of landscape justice allow for an understanding of how 'the dialectic of distribution, participation ... and recognition' (Mels, 2016, p 422) is materialised in particular places, at particular times, and what that means for future transformation.

Normative theories of justice and landscape justice theorising: realising the potential

As indicated, few landscape theorists have heeded Henderson's call from 20 years ago to directly and deeply engage with justice philosophy. Geographers and others have tended to work with an implied sense of 'justice as fairness' and only occasionally engaged more deeply with Rawls' foundational, liberal ideas. Engagement with feminist theorising, particularly the work of Fraser and Young, has, however, been deeper. Recently, the Capabilities Approach has attracted the attention of some landscape scholars, especially those also concerned with energy justice (for example, Mason and

Milbourne, 2019). But, for the most part, this has been gestural: pointing to the value of the Capabilities Approach rather than directly developing it in relation to landscape (as produced space, representation or way of seeing).

There is, however, great potential in finally heeding Henderson's call. In particular, theories of landscape production, landscape representation, access to landscape and use of the landscape could benefit – as the preceding discussion has made clear – from direct engagement with:

- Cosmopolitan theories of *hospitality*, wherein the rights of strangers and others are accorded strong significance, which might offer avenues for broadening the subjects who possess the right to landscape (as Egoz would wish) as well as for developing such rights in a much more scalar-sensitive way.
- Feminist theories of *responsibility* and especially Young's (2011) unfortunately not fully fleshed-out theory of *responsibility for justice*, wherein the *differentiated* responsibilities of distended publics for the perpetuation of exploitative or oppressive social and economic structures are accounted for.
- Marxian/radical theories of justice such as Rainer Forst's (2014; 2017) theory of the *right to justification*, wherein a foundational basis for justice is found in every person's right to *count* in the production and distribution of goods together with their right to have what is done to them in the name of production and distribution *justified*; for Forst, this right to justification is the basis of an *emancipatory* form of justice which might begin to answer Henderson's further call for understanding the role of landscape in offering a 'share of the good life'.
- Further Marxist *theories of exploitation, alienation, and species being*, wherein landscape production (within capitalism) is understood as an inevitably exploitative process, the history of capitalist landscapes is a history of alienation (as Cosgrove argued), and where the benefits that arise out of the making and maintaining landscapes, including profit and the accumulation of capital, is inevitably unevenly distributed, all of which suggest that struggle towards Forst's emancipatory form of justice (one that supports our *species being*) must be fully cognisant of the material, political-economic determinants of the kind of everyday life both J.B. Jackson and George Henderson were concerned with.
- Post-structuralist/feminist theories of *epistemic injustice*, such as that developed by Miranda Fricker (2007; 2013), which offer something like the flipside to Forst's right to justification by focusing not on the right to know and therefore the right to be fully human, but on the *right to be heard and understood*, and therefore to be fully human: this right might very well require significant material transformation as well as epistemic transformation and thus implicate the need to transform systems of

landscape production and maintenance; on this front, the ELC is lacking since it makes no effort to understand the epistemic conditions necessarily at the heart of its processes of participation or even at the heart of its key phrase 'as perceived by people', since that merely raises the question: which people?

While work on each of these fronts would be valuable in and of itself, of even greater value would be their synthesis, difficult as that inevitably will be given their different, not always fully compatible, epistemologies (as Part I of this book made clear).

Such a synthesis would move not only theories of, but efforts to implement, 'justice as fairness', the great liberal ideal, away from the common-sensical and gestural and into the realm of concerted political practice, especially practice aimed at inducing necessary transformations in the mode and relations of landscape production. For it is no stretch at all to suggest that the landscape – as built, physical form (morphology) and also as a mode of representation – is a crucial, indispensable part of what Rawls (1971) defined as the 'basic structure': the institutions and relations without which society itself (and thus human life) is impossible. Landscapes are humanly transformed environments (produced both through our reworking of nature and our reworking of ourselves) and as they are made, they create the conditions of possibility for further development and change. Social relations cannot be changed if their fundamental substrate – the landscape – is not also changed. In this sense landscape is as much a part of the 'basic structure' as is the family (another crucial oversight by Rawls: see the discussion of Susan Moller Okin's work in Chapter 4) and thus an 'object of justice', in Jaggar's helpful typology (see the Introduction).

More specifically, development in each of these areas, and their synthesis, would help clarify how a *substantively just* landscape is one defined not only by a *just distribution* of goods and bads across the landscape, but one hosting a polity in which all have the right to be involved in the *making* of the landscape under non- (or less) exploitative conditions. This polity is itself not (only) locally defined, but extends across scales (cosmopolitan justice) and includes (or should be made to include) excluded and marginalised others (right to justification/epistemic injustice). Since landscape is an object of justice, then *procedurally just* processes for planning, making and preserving landscape which allow for the recognition and representation of the excluded, oppressed and alienated need to be developed, assuring that any such procedures are not a priori dominated by the rule of experts. Only such a substantively and procedurally just landscape – one that minimises epistemic injustice and maximises the right to justification; that offers the possibility to supersede the forces of alienation while being hospitable; that promotes the advancement of species being instead of the accumulation of capital in ever-few hands; and that justly apportions responsibility for harms that will inevitably persist – can ever also be 'beautiful' in the way that Jackson and Henderson urge.

References

Aristotle (1934) *The Nicomachean Ethics*. Cambridge, MA: Harvard University Press.

Barnett, C. (2017) *The Priority of Injustice: Locating Democracy in Political Theory*. Athens, GA: University of Georgia Press.

Berger, J. (1972) *Ways of Seeing*. Harmondsworth, UK: Penguin.

Bruun, M. (2016) How and why was the European Landscape Convention conceived. In K. Jørgensen, M. Clemetsen, K. Halvorsen Thorén and T. Richardson (eds), *Mainstreaming Landscape through the European Landscape Convention*. London: Routledge, pp 5–12.

Cosgrove, D. (1984) *Social Formation and Symbolic Landscape*. London: Croom Helm.

Cosgrove, D. (1985) Prospect, perspective and the evolution of the landscape idea. *Transactions of the Institute of British Geographers*, 10(1), 45–62.

Dalglish, C., Leslie, A., Brophy, K. and Macgregor, G. (2018) Justice, development, and the land: The social context of Scotland's energy transition. *Landscape Research*, 43(4), 517–528.

Daniels, S. (1993) *Fields of Vision: Landscape Imagery and National Identity in England and the United States*. Princeton, NJ: Princeton University Press.

Déjeant-Pons, M. (2006) The European Landscape Convention. *Landscape Research*, 31, 363–384.

Egoz, S. and De Nardi, A. (2017) Defining landscape justice: The role of landscape in supporting wellbeing of migrants, a literature review. *Landscape Research*, 42, 74–89.

Egoz, S., Makhzoumi, J. and Pungetti, G. (eds) (2011) *The Right to Landscape: Contesting Landscape and Human Rights*. Farnham, UK: Ashgate.

Egoz, S., Jørgensen, K. and Ruggeri, D. (eds) (2018) *Defining Landscape Democracy: A Path to Spatial Justice*. London, UK: Elgar.

Fields, G. (2017) *Enclosure: Palestinian Landscapes in a Historical Mirror*. Berkeley, CA: University of California Press.

Forst, R. (2014) *Justification and Critique: Towards a Critical Theory of Politics*. Bristol, UK: Polity.

Forst, R. (2017) *Normativity and Power: Analyzing Social Orders of Justification*. Oxford, UK: Oxford University Press.

Fraser, N. (1997) *Justice Interruptus*. New York: Routledge.

Fraser, N. (2008) *The Scales of Justice: Reimagining Political Space in a Globalizing World*. New York: Columbia University Press.

Fricker, M. (2007) *Epistemic Injustice: Power and the Ethics of Knowing*. Oxford, UK: Oxford University Press.

Fricker, M. (2013) Epistemic justice as a condition of political freedom. *Synthese*, 190(7), 1317–1332.

Hall, S. (ed) (1997) *Representation: Cultural Representations and Signifying Practices*. London: SAGE.

Harvey, D. (1982) *The Limits to Capital*. Chicago, IL: University of Chicago Press.

Henderson, G. (2003) What else we talk about when we talk about landscape: For a return to the social imagination. In C. Wilson and P. rowth (eds), *Everyday America: Cultural Landscape Studies after J.B. Jackson*. Berkeley, CA: University of California Press, pp 178–198.

Jackson, J.B. (1984) *Discovering the Vernacular Landscape*. New Haven, CT: Yale University Press.

Jones, M. (2009) The European Landscape Convention and the question of public participation. In K. Olwig and D. Mitchell (eds), *Justice, Power, and the Political Landscape*. London: Routledge, pp 231–251.

Jones, M. (2016) Landscape democracy and participation in a European perspective. In K. Jøregesen, M. Clementsen, K. Holversen Thorén and T. Richardson (eds), *Mainstreaming Landscape through the European Landscape Convention*. London: Routledge, pp 119–128.

Jorgensen, A. (2016) Editorial: 2016: Landscape justice in an anniversary year. *Landscape Research*, 41(1), 1–6.

Lewis, P. (1979) Axioms for reading the landscape: Some guides to the American scene. In D. Meinig (ed), *The Interpretation of Ordinary Landscapes*. New York: Oxford University Press, pp 11–32.

Mason, K. and Milbourne, P. (2019) Constructing a 'landscape justice' for windfarm development: The case of Nant Y Moch, Wales. *Geoforum*, 53, 104–115.

Matless, D. (1998) *Landscape and Englishness*. London: Reaktion Books.

Meinig, D. (ed) (1979) *The Interpretation of Ordinary Landscapes*. New York: Oxford University Press.

Mels, T. (2016) The trouble with representation: Landscape and environmental justice. *Landscape Research*, 41(4), 417–424.

Mels, T. and Mitchell, D. (2013) Landscape and justice. In J. Duncan, N. Johnson and R. Schein (eds), *A Companion to Cultural Geography*. Oxford, UK: Blackwell, pp 209–224.

Mitchell, D. (1996) *The Lie of the Land: Migrant Workers and the California Landscape*. Minneapolis, MN: University of Minnesota Press.

Mitchell, D. (2003a) Cultural landscapes: Just landscapes or landscapes of justice? *Progress in Human Geography*, 27(6), 787–796.

Mitchell, D. (2003b) *The Right to the City: Social Justice and the Fight for Public Space*. New York: Guilford.

Mitchell, D. (2008) New axioms for reading the landscape: Paying attention to political economy and social justice. In J. Wescoat and D. Johnston (eds), *Political Economies of Landscape Change*. Dordrecht, The Netherlands: Springer, pp 29–50.

Mitchell, D. (2012) *They Saved the Crops: Labor, Landscape and the Struggle over Industrial Farming in Bracero-Era California*. Athens, GA: University of Georgia Press.

Nye, D. (2010) *When the Lights Go Out: A History of Blackouts*. Cambridge, MA: MIT Press.

O'Connor, J. (1998) *Natural Causes: Essays in Ecological Marxism*. New York: Guilford.

Olwig, K. (1993) Sexual cosmology: Nation and landscape at the interstices of nature and culture: Or, what does landscape really mean? In B. Bender (ed), *Landscape: Politics and Perspectives*. Oxford, UK: Oxford University Press, pp 307–343.

Olwig, K. (1996) Recovering the substantive nature of landscape. *Annals of the Association of American Geographers*, 86(4), 630–653.

Olwig, K. (2002) *Landscape, Nature, and the Body Politic: From Britain's Renaissance to America's New World*. Madison, WI: University of Wisconsin Press.

Olwig, K. (2009) The practice of landscape 'conventions' and the just landscape: The case of the European Landscape Convention. In K. Olwig and D. Mitchell (eds), *Justice, Power, and the Political Landscape*. London: Routledge, pp 197–212.

Olwig, K. (2019) *The Meanings of Landscape*. London: Routledge.

Rawls, J. (1971) *A Theory of Justice*. Cambridge, MA: Harvard University Press.

Sauer, C. (1925) The morphology of landscape. In J. Leighly (ed) (1963) *Land and Life: The Writings of Carl Ortwin Sauer*. Berkeley, CA: University of California Press, pp 315–350.

Schein, R. (1997) The place of landscape: A conceptual framework for interpreting an American scene. *Annals of the Association of American Geographers*, 87(4), 660–680.

Schein, R. (ed) (2006) *Landscape and Race in the United States*. New York: Routledge.

Setten, G. (2004) The habitus, the rule and the moral landscape. *Cultural Geographies [Ecumene]*, 11(4), 389–415.

Setten, G. (2020) Moral landscapes. In *International Encyclopedia of Human Geography*, 2nd edn. Oxford, UK: Elsevier, pp 193–198.

Setten, G. and Brown, K. (2009) Cultural geography: Moral landscapes. In *International Encyclopedia of Human Geography*, 1st edn. Oxford, UK: Elsevier, vol 7, pp 191–195.

Setten, G. and Brown, K. (2013) Landscape and social justice. In P. Howard, I. Thompson and E. Waterton (eds), *The Routledge Companion to Landscape Studies*. London: Routledge, pp 243–252.

Torres Garcia, M., Ghislanzoni, M. and Trujillo Camora, M. (2020) The disappearance of public paths in Spain and its impact on landscape justice. *Landscape Research*, 45(5), 615–626.

Wall, E. and Waterman, T. (eds) (2019) *Landscape and Agency: Critical Essays*. London: Routledge.

Wylie, J. (2007) *Landscape*. Oxford, UK: Blackwell.

Young, I.M. (2000) *Inclusion and Democracy*. New York: Oxford University Press.

Young, I.M. (2011) *Responsibility for Justice*. New York: Oxford University Press.

14

Intergenerational Justice

Johanna Ohlsson and Tracey Skillington

Introduction

Justice in the scholarly literature has traditionally and most commonly been approached as an issue, ideal, relation and/or process of change that affects different cohorts of presently living generations (that is, intragenerational justice). Chiefly, these relate to questions of how goods, burdens, responsibilities and harms ought to be distributed among peoples that co-exist. This focus has, to a large extent, been informed and shaped by historical debates on justice where issues of responsibility towards future others have not always carried the same importance. Yet by now, it is well established that many of the activities we engage in during our lifetime will have profound effects on the welfare and composition of generations to come. Contemporary ways of living and acting trigger a need to consider the relevance of an intergenerational justice framework – covering justice and injustices between generations.

Reflecting on these issues, thinkers such as Stephen Gardiner (2001, pp 401–402), Brian Barry (2003) and Axel Gosseries (2009) point to modern lifestyles (and not necessarily population growth) as the biggest threat posed to the future of humanity. Others stress the need to control rates of population growth and, with that, rates of depletion of the Earth's limited natural resource reserves. The emergence of a more structured debate on future ethics and the welfare of generations to come is one that has been influenced by several strands of thinking. Classic philosophers, such as Plato (see, for instance, Lane, 2012), spoke of future generations as a 'direct outgrowth of the present' and noted the 'pious duty' of present generations to 'rear children that will hand on the torch of life from generation to generation' (Hausheer, 1929, p 214). Some have explored Abrahamic as well as Asian non-theistic roots of a principle of intergenerational equity (for instance, see

Weiss, 2021), while others highlight what is distinctive to intergenerational ethics in the age of global climate change (Gardiner, 2010).

The latter strand of thinking does not dismiss the relevance of the ancient theological roots of intergenerational justice but instead chooses to showcase discussions that emerged particularly from the period around the publication of the Brundtland Report in 1987 by the World Commission on Environment and Development. Building on the work of Hans Jonas (1980), Joerg Chet Tremmel, for instance, highlights how the peculiarities of the contemporary ecological age are characterised by a high ability to throw global ecosystems off balance and exhaust what were previously thought to be unlimited reserves of planetary resources (Tremmel, 2009, pp 1–2). With this realisation comes formidable responsibilities to assist those who suffer as a consequence of these actions. Regardless of its historical genealogy, intergenerational justice has become a key concern of more contemporary policy and research communities in several disciplines. Most frequently, it is explored in relation to questions surrounding the causes and effects of global climate change, ongoing economic and social development, sustainability, as well as human and environmental rights.[1] A core component of sustainable development discourse is that of intergenerational equity (Gosseries and Mainguy, 2008, p 2; Weiss, 2021). For instance, the seminal Brundtland Report defines sustainable development as 'development that meets the needs of the present generation without compromising the ability of future generations to meet their own needs' (WCED, 1987, p 53).[2] The need for equity between generations is clearly recognised although what intergenerational justice actually means and how it will be asserted is not clarified (Gosseries and Mainguy, 2008). Policy statements to date offer surprisingly little in the way of guidance on what normative aspects of long-term environmental problems should be acted upon and why. In spite of the level of interest in the concept of sustainable development, equity and intergenerational justice, little effort has been made to translate these concepts into workable models of change. We would suggest that this is partly explained by the absence of clear guidelines as to how these concepts relate to everyday practices and justice reasoning.

Current interest in intergenerational justice and associated concepts is partly explained by the fact that there is no longer a sound basis for denying scientific expert claims that we are on the verge of large-scale ecological breakdown (Oslo Principles on Global Climate Change Obligations; Expert Group on Global Climate Obligations, 2015) and, therefore, must act urgently to prevent disaster and fulfil sustainable development goals. A number of questions are raised by the urgency of this situation. For instance, what obligations are owed to generations who will be most adversely affected by these changes? How can greater equity be established between distant generations? Do future generations have rights that must be respected? If so, what are these rights and how can they be applied, especially in relation to people not yet born?

These questions relate to various ethical, social, political and legal aspects of relations between generations that will be considered in more detail in the following sections. A number of themes are raised consistently throughout the literature on intergenerational justice, and while only some are documented here due to limitations of space, these tend to be themes that are evoked most regularly.

Reoccurring themes in debate on intergenerational justice: issues and critique

Intergenerational or intragenerational justice?

While intergenerational justice relates to justice relations between generations across time, intragenerational justice refers to justice relations among persons living today (Weiss, 2021; Kotlikoff, 2017). One of the key questions for intergenerational justice is whether peoples living today and those living in the future have equal rights to limited natural resources. The fact that stocks of essential resources are declining rapidly means that even if present and future generations are recognised in principle as possessing equal rights to a safe and clean environment, they do not enjoy equal capacity to realise those rights, or have those rights respected. Conflict emerges between the objectives of intragenerational and intergenerational justice when a relationship of rivalry emerges between societal priorities. For instance, key conservation measures (reforestation, rewilding, peatlands preservation, and so on) and efforts to eradicate global poverty (extreme poverty globally rose in 2020 for the first time since the late 1990s [UN, 2021]). Alleviating poverty by increasing access to ecosystem services in the present can lead to a degradation of environmental conditions in the long term, thus reducing availability for future generations. The challenge, therefore, is to achieve a fair balance between priorities, one that recognises the rights of all generations to a portion of essential resources (as a matter of allocated justice) and in doing so, ensures an equilibrium is established between intragenerational and intergenerational justice demands.

There is always a degree of overlap between the demands of co-existing generations, such as the youth and the elderly of today but, also, a degree of difference depending on 'generational location' (shared experiences, cultural interests, ideas and events among a generation [see Mannheim, 1952 (1923)]). Factoring into this equation the needs of distant generations also requires foreplanning and a 'future proofing' of all policy areas in ways that potentially may slow rates of economic and social development in the medium term but improve prospects for sustainable development in the long term.

Time and temporality

The concepts of time and temporality are central to debates on justice between generations (White, 2017).[3] Bound by the limits imposed by birth and death, currently living populations are no more than 'temporary custodians of the planet' (Barry, 2003) who, in being part of a long lineage of custodians, can either 'do a better or a worse job than their successors' (Barry, 2003, p 1) in preserving a sustainable environment. When addressing the role of time in defining our role and place in planetary and social time, Jonathan White (2017) introduces the concept of 'timescape'. Timescapes are said to offer a particular way of representing and conceptualising the contours of time by helping to fix 'its structure, units and scale' (White, 2017, p 764, drawing on Adam, 1998). White argues that a timescape helps to organise perceptions and imagination when it comes to time and temporality just as landscape does in relation to our surroundings. In this way, structure, form and context are accorded to our experiences of time and to what we know about the past and present, but also an unknown future. White clarifies that 'the most far-reaching effects of climate change lie still some decades away, yet that present-day choices will be critical for how those effects play out'. Timescapes, therefore, offer a useful way of visualising ecologically conditioned experiences of time and our relationship to it (White, 2017, p 764).

However, what still remains dominant in mainstream policy and social thinking are linear timescapes. In the case of the latter, time is ordered according to a linear succession of a before and after, between a present-past, a present-present and a present-future, or between 'real time' and 'deferred time' (of climate disaster). From an intergenerational justice perspective, there are several reasons to doubt the validity of this timescape and the way it orders time as no more than varieties of the present, particularly its separation of present reality from everything supposedly opposed to, or beyond, it: absent or non-present future generations.

Linear time allows us to think that we do not owe our lives to those who gave birth to and nurtured us or enabled our democratic freedom (through revolution), or accept that our death does not sever the continuity of life. Linear time affords thinking about various generations' occupation of time as separate. It supports the fantasy of a contemporary life that matricidally gave birth to itself (Derrida, 2013), a life freed from co-constitutive relations to other generations, past and future, and, therefore, freed of duties of care to temporally distant others. Intergenerational justice scholars point to the dangers inherent in such reasoning. Life lives not only generationally and socially in the present but also environmentally in the 'deep time' structures of planetary existence (Chakrabarty, 2015, p 179). Inheritance in this context comes with the promise to support the continuity of life. Using the language of hospitality, Jacques Derrida (1999) sees the 'welcome' granted to temporal newcomers in this instance (that is, future generations) as necessarily a response to a prior welcome given by preceding others.[4] Almost in defiance of these forms of indirect reciprocity connecting generations together, capitalism pursues short-term agendas that serve equally short-term interests. Similarly, democratic culture today reflects what Dennis F. Thompson (2010) refers to as 'presentism', where long-term risks are neglected in favour of short-term goals. There are certain characteristics of democracy, according to Thompson, that encourage 'myopic' reasoning about relations of justice, most notably a need to satisfy the demands and needs of currently voting citizens often at the expense of those that come later. A major reason why this bias towards the present citizenry prevails is the way political power in democratic electoral politics is subject to temporal limits. Rulers exercise political power only for a limited period of time. The tendency, therefore, is to prioritise the interests of present voting publics to maintain their electoral support base. More widely within society, Gábor Bartus (2021, p 266) notes how 'most personal, business and community decisions continue to be made based purely on the desire to achieve maximum short-term benefits'. Presentisms thus continue to tilt legal, political, economic and everyday social decision making towards short-term goals, and negative long-term consequences.

In an effort to address these issues head on, future justice campaigners promote a model of justice and responsibilities towards future generations that recognises the validity of nonlinear time. That is, a timescape that is not identical to, or contemporaneous with, the 'now time' of the present (Skillington, 2015) but references both past wrongdoing and future potentials for reform in the present (World Future Council, 2022). The aim is to shift the focus away from present life as 'my life' to the possibility of making multiple lives across time matter. Justice in this instance cannot be defined in terms of the interests and demands of any one generation or framed exclusively in terms of fixed or unchanging rules and standards of what is just. There is a need for some degree of contingency in light of the growing instability of social and environmental systems (Hendlin, 2014, p 1). This may be seen as a more pragmatic approach to intergenerational justice based on the assumption that the circumstances of ecological and social life can change quite dramatically in the era of advanced climate change. Even so, predictions about the far future are less likely to change in comparison to expectations regarding the near future, thereby making the need for a longer-range perspective on relations of intergenerational justice more obvious (Cowen and Parfit, 1992, p 148). Such arguments may be read as interventions into debates on the uncertainties of climate change and a reappraisal of the time span of intergenerational justice grounded in the belief that we cannot know for certain what will happen in the future, or what will be the long-term outcomes of current decision-making.

The measurability of intergenerational justice: equity, equality or fairness?

Equality is widely considered the chief value of distributive justice (Bidadanure, 2021a, 2021b) but not necessarily that of intergenerational justice. In this setting, 'justice as fairness' (Rawls, 1958, p 164) is more likely to prevail in the interests of imposing a balance between the competing resource interests of the present versus the future. An equal distribution of finite resources across generations is unlikely. What is more likely to be asserted is a difference principle justifying a system of resource distribution that differentially rewards those living in the present in ways that is also minimally advantageous to peoples of the future (beneficiaries of economic development). However, when viewed in terms of overall well-being, basic needs and sufficiency (see, for example, Page, 2007; Tremmel, 2009), the long-term impacts of this system of justice disadvantage all. A collectively relevant justice is not advanced by short-term benefits accrued only by a few and where the situation of the majority is disimproved. The intuitive idea of justice as fairness is that everyone's well-being is improved under a scheme of cooperation and reasonableness without which no one could have a satisfactory quality of life. The division of resource advantages should be such as to secure the willing cooperation of all, including those less well situated (younger and future generations). Yet this is only possible if reasonable and responsible conditions are proposed for the preservation and fair distribution of essential, life-supporting resources. However, this is far from the reality that currently prevails. Already, two-thirds of the global carbon allowance have been burned and if human activities around the globe continue to produce CO_2 at current rates of expenditure, the remaining carbon budget, scientists estimate, will be depleted in little more than a decade (Matthews and Tokarska, 2021). The life chances of all now come to be placed in serious jeopardy. Inequality and denial rather than fairness and reasonableness would appear to be the dominant operative principles. The question then is how is this scenario rationalised?

The non-identity challenge

In *Reasons and Persons* (1984) and 'Energy Policy and the Further Future' (2010, p 116), Derek Parfit assesses whether current scenarios of accelerating natural resource depletion, in producing outcomes that are significantly worse than what was predicted 30 years ago, can be considered unfair. Initially, Parfit's (1984) reasoning was that current resource consumption rates, while generating significant levels of pollution and contributing to global warming, could not be said to harm unborn generations. Parfit's primary reason for making this argument was that future peoples are subjects who at this point in history lack a specific identity (that is, the non-identity problem). Parfit based his argument on a number of factors, including the impossibility of determining at this stage the genetic composition and adaptive capacities of future peoples, both of which are heavily influenced by societal events and the resource choices of would-be parents. In their general capacity to raise standards of living, carbon-intensive energy policies cannot, Parfit argues, be said to harm future generations who would not, in all likelihood, have been brought into existence without the societal improvements generated by carbon pollution. As long as the harms generated by fossil fuel consumption continue to be seen as giving rise to general societal benefits that outweigh diffuse harms, carbon-intensive energy choice will, in all likelihood, Parfit (2010) claims, continue to find support.

The dominance of a 'person affecting principle' (Parfit, 2010, p 119) in institutional assessments of the impacts of increasing carbon pollution means that 'the compensatory benefits' of highly polluting economies, such as those sustained by fossil fuel industries (Parfit, 2010, p 117) will continue to be prioritised over evaluations of their harm. In later writings, Parfit (2010, p 120) points to the pervasive yet deeply problematic nature of this reasoning, noting its failure to address expanding ecological problems with 'wide person affecting' consequences. Not only does this reasoning hinder the further development of a truly sustainable, long-term approach to natural resource management, but it also limits prospects for the realisation of more basic liberal understandings of what a fair system of resource distribution requires – that 'enough and as good' is left for those that follow (Locke's proviso).

The fact that the identity of those that will follow in the future cannot be specified at this point in time (that is, generations not yet born) is of less relevance than the moral duty to conserve essential resources. Restraint in the interests of this wider purpose (conservation) has always been an essential precondition of liberal expressions of justice between generations. For instance, John Rawls' (1971, p 289) principle of just savings formulates restraint as an important 'internal' dimension of liberal democratic approaches to long-term sustainability (Wissenburg, 1999, p 198). However, the understanding also is that 'just savings' be limited to proximate, overlapping generations, rather than temporally distant ones, making the viability of long-term and more globally relevant sustainable development plans less certain.

One political theorist sceptical of proposals to extend principles of justice to future distant others is Terence Ball (2008). Ball highlights how 'the concepts constitutive of our political practices – including "justice" itself – have historically mutable meanings' due mainly to the fact that the circumstances of justice continue to evolve. For such reasons, a transhistorical understanding of justice, according to Ball, is neither practically nor morally justified, especially as it is impossible to know with certainty at this point in time how our actions will affect peoples in the future, or even how such peoples will adapt to harsher ecological conditions, or even if our actions will, in fact, limit options available to them (Howarth, 1992; Ball, 2008, p 321).

A non–identity argument is also evident in arguments in favour of population control (Thompson, 2009). Here little or no attention is given to the character of would-be future peoples. Rather, emphasis is placed on the need to preserve sufficient resources to sustain ongoing economic development. Unsustainable rates of population growth are noted: 'If humans continue to reproduce as predicted, there will be a population of 9.7 billion by 2050 and 10.9 billion by 2100' (Conly, 2021, p 1). What the lives of future others might look like is hence uncertain at this point. However, given current increases in rates of destruction of the Earth's resources, it is likely is that everyday life will be heavily constrained by 'shortages of everything, especially food and drinkable water' (Conly, 2021, p 1). Grave levels of resource deprivation thus prove to be the primary legitimatising mechanism used to support a policy of population control in this instance.

Rights eligibility: the rights of future generations

Whether a correspondence can actually be established between legal rights that exist now and persons who may or may not exist in the future (depending on circumstances) is an issue that provokes considerable debate today. For some, future peoples' non–identity proves too big an obstacle to ever make a rights approach viable. From this perspective, rights to finite resources cannot be imposed if the bearer of these rights is not in a position to exercise them (not yet born). How might rights to limited resources such as a clean and safe atmosphere, for instance, be enforced if the rights holder and obligation bearer do not co-exist? For sceptics, a certain minimum degree of correlativity between parties is necessary if rights are to be actualised in a manner that is meaningful. For these authors, the likely future existence of humans is not sufficient to grant rights and privileges to the same when they do not and may never exist in the years ahead. For instance, if environmental conditions eventually do not support human flourishing, can we meaningfully claim that nonexistent future peoples have been harmed by our failure to preserve a sustainable planet? Further, if future peoples, by virtue of their nonexistence, cannot be harmed by climate change, how can we legitimately claim that they have rights? 'To have' rights is to possess properties to which one can claim a legitimate right. 'What properties do future persons possess at this time?', critics ask.

What is proposed as an alternative is a weak consideration of the interests of future peoples (that is, 'duties owed' to future others) but not necessarily rights. The assumption in this instance is that one can demonstrate how current rates of depletion of fossil fuels, forests, fish stocks and arable lands, in gravely affecting future supplies, affect only the interests of future generations but not necessarily their rights to health, development, a safe environment or freedom from want.

For those in favour of extending rights to future generations, our capacity to predict more accurately future environmental scenarios makes a long-term perspective on human rights eligibility more reasonable. The deep time of ongoing ecological destruction necessitates a critical re-evaluation of relevant contexts for the application of principles of justice across space and time. The equal and inalienable rights of all members of 'the human family' are inscribed in various international law (Universal Declaration of Human Rights; United Nations General Assembly, 1948). More recently, at the Human Rights Council, member states recognised future generations as relevant members of the human family. In October 2021 a resolution recognised the right to a healthy environment and recognised 'further that environmental degradation, climate change and unsustainable development constitute some of the most pressing and serious threats to the ability of present and future generations to enjoy human rights, including the right to life' (HRC Resolution 48/13).

From this perspective, rights ought to be interpreted in ways that transcend the temporal frame of the present and include multiple generations. We are reminded of duties owed to rights holders who are absent at this point in human history but are still defined as relevant subjects of rights. Both the UNESCO Declaration on the Responsibilities of the Present Generations Towards Future Generations (Articles 4 and 5) and the Rio Declaration (Principle 3) offer clarification as to why the welfare and rights of future peoples must be protected. Just as geographical location is thought to have no moral relevance (in principle) to the application of universal human rights, equally, location in time is thought to not always provide sufficiently rational grounds for dismissing all claims that future peoples possess rights. Without temporal or geographical specification, universal rights to liberty, health or development are defined behind a 'veil of ignorance' (Rawls, 1971), that is, without knowledge of the specific circumstances or characteristics of relevant parties to these rights (for example race, gender, age, nationality or social position). Of relevance here are both the individual rights of 'identifiable' living persons to a sustainable environment and the collective rights of those whose identity may not be clearly determinable at this point in time (that is, future generations) but whose need for basic life-enabling resources is able to be determined. The possibility of grounding human rights intergenerationally and formulating justice in deeper, nonlinear temporal terms is therefore something that could be said to be articulated in law already (albeit largely implicitly at present).

References to future generations and responsible use of natural resources are also evident in a significant number of state constitutions. Article 225 of Brazil's constitution, for instance, refers to duties to 'defend and preserve the environment for present and future generations', while the constitution of South Africa (Article 24) affirms every individual's 'right … to have the environment protected for the benefit of present and future generations'. Here references tend to be either general provisions for the protection of future peoples, or more specific references to the natural environment, prone as it is to intergenerational misconduct. While some constitutional provisions focus on the rights of each citizen to environmental protection, others specify the right of every person to an environment conducive to health and emphasise that 'this right will be safeguarded for future generations as well'. Others still focus on the responsibilities of the state to act as a guardian of the resource commons. Article 20a of the German Constitution, for example, defines the role of the state as protecting 'the natural living conditions' of all, including 'future generations'. Collectively, such legislation is an important acknowledgement of legal obligations to preserve opportunities for peoples to survive a deliberate and knowing exhaustion of the Earth's finite natural resource reserves. Second, it validates efforts to prohibit practices that disadvantage or harm the interests (and potentially rights) of present and future peoples. Third, it acknowledges the 'inherent dignity' and 'equal and inalienable rights of all members of the human family' to a safe, just and peaceful world (Universal Declaration of Human Rights; United Nations General Assembly, 1948).

In dialogue with other justice traditions

The Rio Declaration on Environment and Development (United Nations Conference on Environment and Development Rio de Janeiro, Brazil, 1992) notes how 'the right to development must be fulfilled so as to equitably meet the development and environmental needs of present and future generations' (Principle 3). Similarly, the Convention on Biological Diversity highlights the importance of maintaining the potential of ecosystems to 'meet the needs and aspirations of present and future generations' (Article 2). In this way, issues of inter- and intragenerational justice are combined with an integrated understanding of social, economic and environmental development. Formulating the relationship between justice and development in this manner evokes principles of responsibility, indirect reciprocity and a just distribution of resources. All, in turn, are combined with a Capabilities Approach to needs satisfaction. The latter makes the possibility of intergenerational justice seem more plausible in three ways. First, a Capabilities Approach provides a metric of what basic needs must be met for a minimum threshold of sustainable well-being to be secured into the future (that is, the currency of justice). Second, it identifies specific problems that interfere with meeting this basic threshold ('patterns of (in)justice' [Page, 2007]) and, third, it connects the idea of a fair distribution of advantages among peoples with a recognition of the legitimacy of the rights claims of present and duties owed to future generations to protect conditions that enable sustainable development. This puts the needs of all peoples, now and into the future, more sharply into focus and in doing so, increases prospects for a just transition to sustainable futures. Leaning on legal and moral arguments for its justification, the Brundtland Report proposes a definition of needs as a metric of justice. It makes recommendations to governments on how to use such a metric to move towards greater inter- and intragenerational justice. Most importantly, however, in terms of the discussion here, it views the scope of justice as transgenerationally relevant, as does several UN declarations and international agreements. How far the scope of justice ought to extend across generations, however, remains undecided. Many share Rawls' reservations, for example, about applying a principle of reciprocity across many distant generations. For Rawls, it is unfair to expect present generations to work for the benefit of future hypothetical ones who, in not yet being born, are unable to reciprocate. Implicit in this Rawlsian account of chronological unfairness is the notion of a cut-off point to duties owed to future others.

Asymmetrical duties of care, on the other hand, situate multiple generations in relations of reciprocity that stretch forwards and backwards across time (for example, Generation A gives to Generation B who in turn gives to Generation C), with each seen as complicated in the building of sustainable futures (Skillington, 2019a) and contributing centrally to the good of all. According to this perspective, knowledge of the cumulative effects of environmental destruction and the steady disappearance of safe ecological time compel a more serious effort to take account of the welfare of absent generations. What are evoked in this instance are not only principles of responsibility and just distribution but, also, a cosmopolitan recognition of duties of care (hospitality) owed to future others. Ongoing deteriorations in ecological heritage threaten to 'impoverish' the heritage of all, present and future (UNESCO, 1972) and, therefore, necessitate actions that protect 'our common future' (Bruntland, 1987).

Procedural justice for present and future generations

It may seem sensible to some to limit conceptions of justice to the living on account of the natural limitations of future generations to participate in discussions (typically assumed to be a necessary precondition). For youth campaigners, however, what is more important is that prerequisites of presence and fair representation be extended to include today's youth and citizens of the future. Youths bring to the fore issues of status inequality and unfair exclusions of the basic interests of future generations, including the right to representation in decision-making processes on issues that fundamentally affect prospects for healthy future living (that is, decisions on the pursuit of high-risk energy options). The desire of youth to participate in decision-making and represent also generations to come is motivated by at least three factors:

1. the fact that youth are the face of the future;
2. youth bring fresh perspectives to bear on issues which older generations might not think of, or take seriously; and
3. youth's desire to participate is also motivated by legal rights.

The Convention on the Rights of the Child, for example, provides a clear and solid foundation for the assertion of the rights of youth to participate in decision-making on matters that affect their lives and the right to protection.

A number of developments in recent decades highlight the capacity of existing political decision-making fora and legal frameworks to accommodate a more intergenerationally grounded perspective on environmental imperatives. For instance, the decision of the Maltese government in 1992 to appoint a 'guardian for future generations' to alert wider policy communities of the importance of long-term perspectives on key policy issues. In 2007, the Hungarian Parliament established a Parliamentary Commissioner for Future Generations to 'ensure the protection of the fundamental right to a healthy environment' (Ambrusne, 2010). Similarly, in Wales, a Welsh Commissioner for Future Generations was established by the Welsh Assembly in 2011. Trusteeship of the democratic process for future generations, therefore, is not an entirely new venture. However, models to date have proven to be quite limited. Trustees' powers could be enhanced significantly, especially their legal powers, to ensure recommendations are acted upon and policy changes made (for instance, a fuller integration of posterity impact assessments into all current policy evaluation measures [see Thompson, 2010]).

For more radical thinkers, trustee activities, while an important step forward, are not sufficient in themselves. As trusteeship is currently not based on an acknowledgement of the rights of future generations (legally grounded) but, rather, duties of care owed to the same, measures to extend democratic privileges to generations to come still lack institutional support (World Future Council, 2022). As one of the great virtues of democratic societies, accountability can become a hindrance to intergenerational justice if its powers are used to prioritise the interests (for example, the energy demands) of presently living generations. To ensure democratic virtues are preserved for generations to come, Derrida (2005) calls for democratic rights and even citizenship to be extended to the unborn. As Derrida observes, democracies are always simultaneously a question of 'what is' and 'what is to come'. Defining relevant members of this democratic community requires that the social ontological futurity of time be taken into consideration and democratic relations and commitments fostered between subjects who are living, dead and not yet born (*les vivants et ceux qui doivent encore naître* [see Derrida, 2004]).

Conclusion

The discussion in this chapter notes how the challenge of non-presence poses the biggest hurdle to efforts to strengthen the normative relevance and legal status of future generations. It assesses some of the main arguments put forward by those who claim that presence matters in the allocation of justice, as well as those who refute this claim and instead highlight possibilities for the actualisation of a radical model of trusteeship that protects the rights and interests of present *and* future generations. A second major challenge to intergenerational justice is the rivalry that can emerge between societal priorities – that is, for instance, eliminating poverty and efforts to conserve remaining essential resource reserves.

Whatever model of intergenerational justice prevails, it is one that has to acknowledge the current historical moment as one of unprecedented crisis and unresolved contradictions between competing agendas (economic, geopolitical, ecological). With the possibility of humanity's 'non-future' looming ever large, the boundaries of what is imaginable in justice terms must be stretched to encompass the 'not yet' moment of humanity's existence. If that means assigning privileges to future generations and honouring age-old democratic commitments to a better world, so be it.

Notes

[1] The most common disciplines seem to be sociology, political theory, economics and sustainability studies, as well as political and moral philosophy.

[2] Here, see, for instance, Julian Agyeman's discussions on *just sustainabilities* as a mode of generational thinking, as this is most relevant and helpful for exploring the juncture of sustainability and justice, and the increasingly intersecting goals of social justice and environmental sustainability, also linking it to Chapter 9 on environmental justice, and Chapter 15 on just transitions (Agyeman, 2013).

[3] The concept of temporality has often been interpreted by philosophers as linear, starting in the past, continuing to the present and to the future. However, several scholars have questioned this linear way of thinking, arguing instead for a nonlinear approach, often in relation to modernity (see, for instance, Heidegger, 1927) and perhaps post-modernity. The concept of time has been a theme for philosophical scrutiny for thousands of years, but that is partly a separate discussion, and perhaps not of too much importance for understanding temporality in relation to intergenerational justice, but some awareness of this could be helpful.

[4] This may also be seen as a reciprocal justice across generations, one that is common in much of climate justice theorising (see Chapter 10, and, for instance, Gosseries, 2009; Caney, 2014; Skillington, 2019b; Gardiner, 2021) and campaigning (see, for instance, the Foundation for the Rights of Future Generations). Similarly, it emerges in Indigenous approaches to justice (see Chapter 7, this volume) although in this instance, it tends to

be oriented towards earlier generations (restitution for past wrongdoing and honouring the dead).

References

Adam, B. (1998) *Timescapes of Modernity: The Environment and Invisible Hazards.* London: Routledge.

Agyeman, J. (2013) *Introducing Just Sustainabilities: Policy, Planning, and Practice.* London: Zed Books.

Ambrusne, E.T. (2010) The parliamentary commissioner for future generations of Hungary and his impact. *Intergenerational Justice Review*, 10(1), 18–24.

Ball, T. (2008) The incoherence of intergenerational justice. *Inquiry*, 28(1–4), 321–337.

Barry, B. (2003) Sustainability and intergenerational justice. In A. Dobson (ed), *Fairness and Futurity: Essays on Environmental Sustainability and Social Justice.* Oxford, UK: Oxford University Press, pp 93–117. https://oxford.universitypressscholarship.com/view/10.1093/0198294891.001.0001/acprof-9780198294894-chapter-5

Bartus, G. (2021) Unsustainability as an economic problem. In M. Cordonier Segger, M. Szabó and A. Harrington (eds), *Intergenerational Justice in Sustainable Development Treaty Implementation: Advancing Future Generations Rights through National Institutions.* Cambridge, UK: Cambridge University Press, pp 267–280.

Bidadanure, J. (2021a) *Justice Across Ages: Treating Young and Old as Equals.* Oxford, UK: Oxford University Press.

Bidadanure, J. (2021b) Justice between co-existing generations. In S. Gardiner (ed), *Oxford Handbook of Intergenerational Ethics.* Oxford, UK: Oxford University Press.

Brundtland, G. (1987) *Report of the World Commission on Environment and Development: Our Common Future.* United Nations General Assembly document A/42/427.

Caney, S. (2014) Climate change, intergenerational equity and the social discount rate. *Politics, Philosophy & Economics*, 13(4), 320–342.

Chakrabarty, D. (2015) *The Human Condition in the Anthropocene: The Tanner Lectures in Human Values.* Yale University, February. https://tannerlectures.utah.edu/_resources/documents/a-to-z/c/Chakrabarty%20manuscript.pdf

Conly, S. (2021) The challenge of population, in S. M. Gardiner (ed), *The Oxford Handbook of Intergenerational Ethics*, online edn. Oxford, UK: Oxford Academic.

Cowen, T. and Parfit, D. (1992) Against the social discount rate. In P. Laslett and J.S. Fishkin (eds), *Justice between Age Groups and Generations.* New Haven, CT: Yale University Press, pp 144–161.

Derrida, J. (1999) *Adieu to Emmanuel Levinas*. Stanford, CA: Stanford University Press.

Derrida, J. (2004) *For What Tomorrow ... A Dialogue with Elizabeth Roudinesco*. Stanford, CA: Stanford University Press.

Derrida, J. (2005) *Rogues: Two Essays on Reason*. Stanford, CA: Stanford University Press.

Derrida, J. (2013) The night watch. In A. Mitchell and S. Slote (eds), *Derrida and Joyce: Texts and Contexts*. Albany, NY: SUNY Press, pp 87–108.

Expert Group on Global Climate Obligations (2015) *Oslo Principles on Global Climate Change Obligations*. https://globaljustice.yale.edu/sites/default/files/files/OsloPrinciples.pdf

Gardiner, S. (2001) The real tragedy of the commons. *Philosophy and Public Affairs*, 30(4), 387–416.

Gardiner, S. (2010) A perfect moral storm: Climate change, intergenerational ethics, and the problem of moral corruption. In S. Gardiner, S. Caney, D. Jamieson, H. Shue, P.T.F.S. Caney, P.D.O.E.S. Jamieson and S.R.F.H. Shue (eds), *Climate Ethics: Essential readings*. Oxford, UK: Oxford University Press, pp 87–98.

Gardiner, S. (2021) *The Oxford Handbook of Intergenerational Ethics*. Oxford, UK: Oxford University Press.

Gosseries, A. (2008) Theories of intergenerational justice: A synopsis. *S.A.P.I.EN.S. Surveys and Perspectives Integrating Environment and Society*, 1.1. http://journals.openedition.org/sapiens/165

Gosseries, A. (2009) Three models of intergenerational reciprocity. In A. Gosseries and L.H. Meyer (eds), *Intergenerational Justice*. Oxford, UK: Oxford University Press.

Gosseries, A. and Mainguy, G. (2008) Theories of intergenerational justice: A synopsis. *SAPIENS*, 1(1), 61–71.

Hausheer, H. (1929) Plato's conception of the future as opposed to Spengler's. *The Monist*, 39(2), 204–224.

Heidegger, M. (1927) *Being and Time*. New York: Harper & Row.

Hendlin, Y.H. (2014) The threshold problem in intergenerational justice. *Ethics and the Environment*, 19(2), 1–38.

Howarth, R.B. (1992) Intergenerational justice and the chain of obligation. *Environmental Values*, 1(2), 133–140.

Jonas, H. (1980) *Philosophical Essays*. Chicago, IL: University of Chicago Press.

Kotlikoff, L.J. (2017) Measuring intergenerational justice. *SSOAR*, 3(2), 56–63.

Lane, M. (2012) *Eco-Republic: What the Ancients Can Teach Us about Ethics, Virtue, and Sustainable Living*. Princeton, NJ: Princeton University Press.

Mannheim, K. (1952 [1923]) The problem of generations. In K. Mannheim, *Essays on the Sociology of Knowledge*. London: RKP, pp 276–320.

Matthews, H.D. and Tokarska, K. (2021) New research suggests 1.5C climate target will be out of reach without greener COVID-19 recovery plan, *The Conversation*, January. https://theconversation.com/new-research-suggests-1-5c-climate-target-will-be-out-of-reach-without-greener-covid-19-recovery-plans-151527

Page, E.A. (2007) Intergenerational justice of what: Welfare, resources or capabilities? *Environmental Politics*, 16(3), 453–469.

Parfit, D. (1984) *Reasons and Persons*. Oxford, UK: Oxford University Press.

Parfit, D. (2010) Energy policy and the further future. In S. Gardiner, S. Caney, D. Jamieson, H. Shue and R.K. Pachauri (eds), *Climate Ethics: Essential Readings*. Oxford, UK: Oxford University Press, pp 112–121.

Rawls, J. (1958) Justice as fairness. *The Philosophical Review*, 67(2), 164–194.

Rawls, J. (1971) *A Theory of Justice*. Cambridge, MA: Harvard University Press.

Skillington, T. (2015) Theorizing the Anthropocene. *European Journal of Social Theory*, 18(3), 229–235.

Skillington, T. (2019a) *Climate Change and Intergenerational Justice*. Oxon, UK: Routledge.

Skillington, T. (2019b) Changing perspectives on natural resource heritage, human rights and intergenerational justice. *International Journal of Human Rights*, 23(4), 615–637.

Thompson, D.F. (2010) Representing future generations: Political presentism and democratic trusteeship. *Critical Review of International Social and Political Philosophy*, 13(1), 17–37.

Thompson, J. (2009) *Intergenerational Justice: Rights and Responsibilities in an Intergenerational Polity*. New York: Routledge.

Tremmel, J.C. (2009) *A Theory of Intergenerational Justice*. London: Routledge.

United Nations (UN) (2021) End poverty in all its forms. *UN Sustainable Development Goals*. https://unstats.un.org/sdgs/report/2021/goal-01/

United Nations Conference on Environment and Development Rio de Janeiro, Brazil (1992) *Agenda 21: Programme of Action for Sustainable Development; Rio Declaration on Environment and Development; Statement of Forest Principles*. Rio de Janeiro, Brazil: United Nations Deptartment of Public Information. A/CONF.151/26 (vol. I), 31 ILM 874.

United Nations Educational, Scientific and Cultural Organisation (UNESCO) (1972) *Convention Concerning the Protection of the World Cultural and Natural Heritage*, 16 November.

United Nations General Assembly (1948) *Universal Declaration of Human Rights* (217 [III] A). Paris: United Nations General Assembly.

WCED (World Commission on Environment and Development) (1987) *Our Common Future*. UN document: A742/427.

Weiss, E.B. (2021) Intergenerational Equity, Oxford Public International Law, Max Planck Encyclopedias of International Law, Encyclopedia entry. https://opil.ouplaw.com/display/10.1093/law:epil/9780199231690/law-9780199231690-e1421?prd=MPIL

White, J. (2017) Climate change and the generational timescape. *The Sociological Review*, 65(4), 763–778.

Wissenburg, M. (1999) An extension of the Rawlsian savings principle to liberal theories of justice in general. In A. Dobson (ed), *Fairness and Futurity: Essays on Environmental Sustainability and Social Justice*. Oxford: Oxford University Press, pp 173–198.

World Future Council (2022) *Save the Planet, Protect the Future*. https://www.worldfuturecouncil.org/save-the-planet-protect-the-future-no-excuses-for-inaction-eight-policy-measures-that-governments-should-take/

15

Just Transitions

Darren McCauley

Introduction

Just transition is an emerging concept of global importance for both theory (Weller, 2019; Burke, 2020; Crowe and Li, 2020; Shen et al, 2020) and practice (ILO, 2016; FoE, 2017; Presidency, 2018; European Commission, 2020a). Just transition is defined here as ensuring a fair and equitable process of moving away from fossil fuels and towards the adoption of renewable and low-carbon technologies, disrupting, reconfiguring and usurping the prevailing carbon-intensive global regime. It is a less studied area for justice theorisation in comparison to environmental or climate justice, borrowing similar conceptualisations and frameworks. Just transition has a unique grounding within the trade union movement (Stevis and Felli, 2015; Mayer, 2018). This means at the centre of its usefulness is its critical reflection on workers' rights. As explored in more detail in this chapter, it has two distinct phases of historical development, one from the 1980s when it focused almost uniquely on employment to a resurgence in both theory and practice. Since 2015, its gaze expanded to include both workers' rights as well as an interest in inequalities emerging across the transition away from fossil fuels.

The contemporary use of the just transition concept is now firmly rooted in its ability to encourage critical reflection on the societal and environmental implications of the transition away from fossil fuels (McCauley and Heffron, 2018; Heffron and McCauley, 2019; Cha et al, 2020; Lawrence, 2020). This transition necessarily involves the adoption of renewable and low-carbon industries. Both moving away from fossil fuels and towards renewables are of concern to scholars and practitioners in just transition. Its most recent emergence is driven by international agreements on the need to transition away from fossil fuels, especially since the Paris Agreement of 2015 (UN, 2020). Environmental, climate and the newer area of energy justice tend to avoid explicit reflection on the reality of this transition. Just transition is used increasingly to focus our attention back on this journey.

Overview of approach

Just transition scholars are predominantly concerned with workers' rights. Work in this area involves a wide range of related issues from industrial relations to human resources policies. It tends to emphasise the impacts of lost industries such as coal. The replacement or relocation of industries mean a loss of employment opportunities for affected communities. This brings into play several relevant foci. The first is the nature of what is meant by transition. The restoration of a sense of fairness in affected communities is a second consideration. This has, third, led to scholars considering the wider implications of the transition, directly on workers' rights as well as more indirectly on the community at large.

What is just transition?

The transition is explicitly defined within the context of moving away from fossil fuels. Within this context, scholars argue reflection is needed on employment impacts (Evans, 2007; Swilling et al, 2015; Goddard and Farrely, 2018; Cha et al, 2020; Crowe and Li, 2020). In this way, the term is intimately associated with employment considerations in the transition away from fossil fuels. This presents a high degree of conceptual separation from other existing justice scholarships such as climate or environmental justice, which are less concentrated on the employment aspects of the transition (see Chapters 9 and 10). Energy justice and just transition are, for example, two related but distinct concepts. Energy justice refers to the fair distribution of benefits and costs associated with energy production and consumption, considering social, economic and environmental factors (see Chapter 11). Just transition, on the other hand, is a framework for ensuring that the shift to a low-carbon economy is equitable and inclusive, particularly for workers in industries that are being phased out. While both concepts concern equity in the energy sector, energy justice focuses on broader issues of fairness and access, while just transition deals specifically with the transition to a sustainable future. The past, current and future use of the just transition term must appreciate the centrality of livelihoods and employment to retain its central focus. This should not, however, hamper its conceptual development to include other factors to make the concept more robust and adoptable (Heffron and McCauley, 2018; McCauley and Heffron, 2018). I see this process in action in the section on 'Debates in just transition', in which I investigate critical debates in the field.

The jobs argument central to the concept is focused on the differential impact of the transition across geographical locations. There is much debate about the real impact of the low-carbon transition on employment (Kenfack, 2019; E2, 2020; OECD, 2020). Some recent reports have concluded, for example, that such a transition will have a minimal overall impact on employment in Europe (OECD, 2020). Such reports often neglect to explore the variation geographically in both its direct and indirect impacts on associated sectors of employment in fossil fuel industries. As a result, trade unions have driven the development of the just transition approach globally. The European Trade Union Confederation concluded that 'from a worker's perspective, the transition will profoundly reshape the labour market in ways that create new employment but also in some cases destruction of jobs' (ETUC, 2018). This has meant that scholars have turned their attention to trade union politics and their framing strategies for lobbying institutions (Rueckert, 2018). The ways in which trade unions mobilise, strategise and frame the transition has been a central focus for just transition scholarship (Stevis and Felli, 2015). Its unitary focus on employment has led to its relative marginalisation in justice scholarship when compared to climate and environmental justice, where connections have not yet been made explicit. I argue in line with others (McCauley, 2018; Weller, 2019; Crowe and Li, 2020) that its conceptual widening beyond employment provides an opportunity for more critical reflection on the transition away from fossil fuels.

Conceptualising transitions

The 'transition', or transitions, is a key component of this discussion. It is partly overlooked in existing literature on just transition. I respond to this gap by briefly delving into the dominant literature on transitions to provide some originality to this literature coverage. To understand what a just transition is, or could be, clarity is needed to identify exactly what I mean by 'transition'. The dominant literature in this field is 'social technical systems' or 'the multilevel perspective' (Scrase and Smith, 2009; Mullally and Byrne, 2016; Schot et al, 2016; Sareen and Haarstad, 2018; van Welie and Romijn, 2018). It has guided scholarships around questioning and formulating sustainable transitions. Just transition and just sustainabilities are closely related concepts. Just sustainabilities is a framework that emphasises the importance of integrating social justice and environmental sustainability in decision-making processes (Agyeman, 2013). Just transition, on the other hand, refers to the need for a fair and equitable shift from an unsustainable economy to a more sustainable one. Both concepts share a focus on equity and justice, recognising that marginalised communities often bear the brunt of environmental degradation and economic transitions.

A transition is a journey and the outcome of this journey in this field is sustainability. As this volume is not focused on sustainability as such, I limit its definition to the core agreed Brundtland (1987) definition based on social, environmental and economic sustainability. This approach towards transition provides complexity in terms of the process for achieving this outcome. It conceptualises the transition as a process through the adoption of new technologies or ideas to achieve a change in society. It shares the same concern with temporality as climate justice (see Chapter 10), and to some extent intergenerational justice (Chapter 14). The *process* for this literature on socio-technical systems is equally as important as the outcome, sustainability. A stereotypical example of this approach to transition is the rise of electric vehicles (Zhang, 2014; Yang et al, 2018). They represent a major disruption that takes place through technological adoption, resulting in the reconfiguration of transport behaviours and institutions. It is inherently derived from the consideration of technological concerns with society.

The bottom-up process of technological-based disruptions must be accompanied by reflection on the macro level of the transition. The ultimate goal is to create a system of institutions, ideologies and actors which will establish a regime leading to greater stability and above all sustainability (Calvert et al, 2019). The regime is understood to be currently unsustainable. There, elite actors and institutions are attempting to resist the positive change needed to achieve sustainability. An example of socio-technical systems involves the slow adoption of housing insulation schemes in cities and the hesitancy of the construction industry in adopting the requirement for more sustainable buildings to be constructed (Howden-Chapman et al, 2005).

Fairness, equity and justice have very recently emerged as considerations for this process of transition in socio-technical systems (Vilches et al, 2017). Its focus remains nonetheless on the variety of possible sustainable transitions and the technologies that drive them, rather than a systematic analysis of inequalities (Jenkins et al, 2018). The just transition literature is where a focus on justice is more likely. For just transition, technological disruptors are understood to be the idea of decarbonisation alongside renewable and low-carbon technological adoption. The status quo is currently viewed as fossil fuel interests that resist this change that will need ultimately to be more sustainable.

Debates in just transition

The adoption of just transition by societal actors at all levels is evident – and not without critique. It is both an area of academic interest as well as for practice. Unlike environmental or climate justice, the emergence of just transition in practice is not limited to environmental organisations or even civil society more generally. It is increasingly adopted in national governments, international and regional institutions as well as in the actions and strategies of private organisations. Three critical debates emerge when considering the implications of an increasingly practised and applied just transition concept.

A concept developed by trade unions, for only trade unions?

Trade unions have led the way in promoting the concept of just transition. I see several examples throughout the world where a just transition frame has been used as a campaign tool for national and international trade unions. The first recorded use of the term by trade unions in the 1980s was in the United States (Abraham, 2017). It came in response to the US government's stated policy in 1986 to pursue the closure of coal mines. It was stated that the mining activities for many communities were to be replaced by renewable energy employment opportunities. To develop a sense of collective buying, American trade unions united around the just transition concept as a formal demand to government that each job lost in the coal mine closures would be replaced by renewable energy development (Peerla, 1999). This has actually inspired non-trade union applications, however.

It directly led to a dispute found in many countries across the world between brown and green energy transitions among civil society organisations (Evans and Phelan, 2016). The just transition term was an example of a 'brown' understanding of the transition away from fossil fuels. It prioritised social issues, in this case employment restoration, as the key priority. In the 1990s and early 2000s, a response emerged from environmental organisations which sought to use the term, albeit sparingly, in relation to a green approach to the transition (Agyeman, 2013). This led organisations such as the World Wildlife Fund (WWF) to argue that a just transition is equally about protecting the environment. It was quickly replaced by climate justice as a dominating frame for environmental organisations. It has nonetheless re-emerged among environmental organisations in unison with trade unions. WWF Germany ran a campaign for a just transition, incorporating both environmental and social concerns in relation to the sharp increase in coal industrial activity after the decision to move away from nuclear energy (Fuller and McCauley, 2016). Friends of the Earth Scotland succeeded in driving forward a nationwide campaign on just transition in relation to the decommissioning of oil and gas rigs (FoE, 2017). The just transition frame has therefore moved away from its trade union movement origins towards a more civil society-wide approach where country or region-specific campaigns are run to bring attention to emerging social and environmental inequalities.

The development of just transition as a key concern for civil society organisations has driven more recent attempts at developing national, regional or project-specific commissions throughout the world. The Scottish example is a world-leading embodiment of such a process (Gov, 2018). A national commission for just transition was established in 2018. It involves multiple interests in society from universities to trade unions and environment organisations. These are formalised spaces for multiple stakeholder interaction designed to develop strategies for social and environmental inequalities-based action. Moreover, the International Labour Organization has succeeded in establishing a range of just transition guidelines for companies in their activities in developing nations (ILO, 2016). In summary, the just transition term is actively in use throughout the world on a range of levels, but most importantly involving both social and environmental calls to aim to tackle the inequalities of the transition.

Co-opted by international elites?

The just transition frame has also a separate and unique historical development among international institutions. This has raised the implication that it is now co-opted by international elites with the objective of undermining its original purpose. The success of trade unions, and then environmental organisations, to bring the just transition idea to the fore is critical for explaining its adoption by a range of international institutions. The key point to acknowledge at this stage is that a similar development of the concept as outlined here has not yet taken place as just transition as a concept has been absorbed into international organisations' politics and agendas. It is evident that just transition as an idea remains limited to the job replacement understanding. The first emergence of the term in an international setting was explicitly the Paris Agreement in 2015. A world-leading agreement on climate change involving 188, and now 189, nations, as well as a range of international representation, equally adopted just transition as an objective. At the time, this did not receive much attention. The International Labour Organization has claimed its critical role in ensuring that a just transition was mentioned in the Paris Agreement as an explicit objective to ensure that any move towards a post-carbon or decarbonised world involves the replacement of lost jobs. In a sense, an international reboot of the concept has taken place.

The adoption of just transition in the Paris Agreement as an objective (UN, 2020) was quickly followed up most notably in the G7 annual meeting of national leaders in 2018 (G7, 2018). It explicitly stated that the just transition is defined as ensuring a fair and equitable approach to the transition away from fossil fuels for all workers. The just transition tone is placed firmly within its dominant economic perspective between employers and employees, driven by capitalist market understandings of inequality. The now famous Silesia declaration was signed by over 60 countries in the world, committing their nations to accord special attention to the loss of coal mining and associated activities (Presidency, 2018). The formalisation of the 1980s understanding of just transition is now firmly in place. The International Monetary Fund, United Nations Development Programme and the World Trade Organization all now have just transition strategies that have been re-developed from this emergence in the international community. The real consequences of the just transition adoption, and this understanding, is most easily understood when assessing the associated financial mechanisms.

It is unfortunate that just transition has not undergone the same maturing process at an international level as it has among civil society organisations. The real-world impact of this reality is played out in a range of national and international financial mechanisms that are emerging. The primary issue is that just transition is a conduit for fossil fuel investment. This is counter to its original intended use. It has become formalised as a support mechanism for fossil fuel areas of the world that are suffering from industrial closures. Germany has set aside a specific mechanism that allocates over €40 billion to coal regions from July 2022 to 2038 (CleanEnergyWire, 2020). The rationale is to fund retraining programmes as well as community initiatives for helping affected workers and their families. A similarly impactful figure of £62 million was announced in June 2020 for affected oil and gas communities in Scotland as part of the just transition commission mentioned previously (Scottish Government, 2020). However, the most notable perpetrator of such an approach to just transition is the European Union (EU). I explore this in some detail in the next section.

Greenwashing through the European Union Green Deal?

The EU Green Deal was established in 2019 and throughout 2020 as the first multinational and international commitment to supporting the transition away from fossil fuels. It commits a much-disputed figure of over €1 trillion towards financing the transition. A central component of this deal is the just transition platform with its associated financial mechanisms. It provides a comprehensive dedicated just transition fund, investing EU and the European Investment Bank public sector loan facility as a triumvirate of ways to invest in European regions (EP, 2020). It is the most sophisticated international collaboration of just transition in the world. It involves the necessary construction of regional and national plans if financing is to be secured in each case. The approach is therefore project driven whereby organisations apply for funding for these three funding opportunities. The main issue emerges as those regions that are most affected are not necessarily the geographical locations where renewable or low-carbon technologies will flourish. In short, there is a clear threat that this investment only serves to embed fossil fuel activities and the carbon-intensive regime further in these regions.

This understanding of just transition is clearly at odds with the more mature elaboration found within civil society. The EU itself recognises this potential threat, commenting in early 2020 that 'the just transition platform will bring together expertise from all relevant commission services to make sure that fossil fuel and carbon-intensive regions have all the information, tools and assistance they need to transform their economies in a fair way' (European Commission, 2020a). It is questionable whether financing such regions should be the priority in following the objectives of the concept. Some have also raised the peculiarity of financing fossil fuel interests at the same time as the Commission is adopting ambitious climate change targets (European Commission, 2020b). Investing in the areas where future renewable technologies are being developed would be a more appropriate financial setup with explicit care taken to environmental impacts. Most recent events have pointed unfortunately to the opposite.

On 16 September 2020, the EU Parliament revised the just transition fund to be explicitly available for natural gas projects. This is a clear example of the dominance of fossil fuel-based interests in the just transition agenda in the EU (WWF, 2020). As one of the three key funding mechanisms under the just transition platform, the just transition fund was explicitly targeted towards cleaning coal-based activities or renewable projects. The revisions taken by the European Parliament have opened the door for gas to be positioned as a so-called transitionary fuel to the detriment of renewable-based investment opportunities. As the just transition platform is at an early stage, the example of Poland is the most relevant at this stage. The most carbon-intensive regions of the EU are in Poland. The European Commission has already identified that it will be the most significant recipient of funds. It is an exemplar of the hypocrisies of the just transition fund. The regions of most significant potential are not the old carbon-intensive communities, with the notable exception of the Lublin basin. Investment is likely to go into the south of the country, where areas of greatest renewable capacity potential remain in the north. An urgent maturation process of the just transition concept at an international institutional level is required to avoid just transition funds being used to further embed carbon-intensive activities.

Just transition: moving beyond the normative application of restorative justice?

The just transition concept lacks the philosophical maturity of more established areas such as environmental justice, making it ripe for further theoretical reflection. It has only recently emerged beyond its trade union campaign motif. I argue that its wide adoption throughout society means that it must, first, embrace a more comprehensive approach to restorative justice and then, second, engage in core justice thinking on distributional and procedural dimensions.

Applying a comprehensive approach to restorative justice

Rather than procedural justice, restorative justice is the central notion that has influenced and been applied to the just transition concept. It is intimately associated with law. It involves deep questioning after a perceived or experienced injustice has taken place (Gibbs, 2009; Preston, 2011). It is sometimes referred to in the literature as corrective justice (McAlinden, 2011). It has been mostly applied to criminal law rather than civil. The application of restorative justice to the transition sheds light on the criminal activities of the fossil fuel industries in exploiting their dominance of the fossil fuel regimes across the world (Goodstein and Butterfield, 2010; Preston, 2011; Lawrence and Åhrén, 2016; Leijten, 2019). I have indeed seen the application of criminal law increasingly with regards to energy companies and their activities. Restorative justice is applied in just transition scholarship as job replacement where restoration means the replacement of lost employment in fossil fuel industries with new employment in the renewable sector.

Just transition scholarship has shed light on the community impacts of job losses throughout predominantly a North American context (Snyder, 2018). The closure of fossil fuel industries has not been geographically met with renewable energy job replacements. This has resulted in existing literature calling for new forms of financial compensation (Zadek, 2019) or the strategic replacement of these industries with new technologies (Pollin and Callaci, 2019). The process of restoration is not simply about direct employment but rather the widespread indirect impacts on a given community. Some just transition literature has emphasised the importance of relocation of fossil fuel industries, rather than their closure, pointing towards the inherent unfairness for communities and without any real benefit to the transition (Altintzis and Busser, 2014; Patterson and Smith, 2016; Healy and Barry, 2017). They explore the initiatives started by local governments or companies that have sought to restore a sense of justice, often considered at local community level to include ideas of environmental remediation from negative energy industry impacts, or the re-establishment of past livelihoods before energy industry employment. Another thread of literature looks at social licence to operate as a framework for restoring justice through building in post-hoc systems of recompense after an industry has left (Hall et al, 2015; Jijelava and Vanclay, 2017). Just transition in this way brings our attention to the tangible and intangible losses incurred by communities when industries move and the ways in which governments seek to alleviate growing senses of injustice.

There is an opportunity to broaden the concept further by addressing the environmental impacts of such processes. This has emerged in recent literature where both indirect and direct employment effects have been connected to observations on broader environmental impacts of fossil industries (Harrahill and Douglas, 2019; Cha et al, 2020; Shen et al, 2020). The inadequacies of post-mining environmental restoration have led to scholars using just transition as a framework to analyse the inequalities experienced by local people in relation to their physical environment (Weller, 2019). Such concerns are normally outlined in environmental and energy justice scholarship (Smith, 2013). The intimate connection with the fossil fuel industries has led trade union-based explorations to consider the wider physical environment implications of closures and relocations. This recent literature demonstrates an opportunity to expand further the remit of just transition. The social impacts of the transition are in this way intimately linked to environmental consequences. A sustainable transition necessitates consideration of both.

Applying distributional and procedural justice in just transition

Spaces of proximity and due process are equally applicable to the just transition frame as I find in climate, energy and environmental justice scholarships. As can be seen, the relocation and geographical sensitivities of fossil fuel industries have been long considered part of the just transition area of research (Abraham, 2017). Proximity in this case offers a new way of understanding what types of inequalities may emerge in the transition, both for social and environmental concerns. The fossil fuel industry structure of centralised energy generation or extractive activities is to be replaced by decentralised smaller infrastructures that do not require the same level of maintenance and therefore direct community employment (Crowe and Li, 2020). This is compounded further by a more geographically limited opportunity for rare earth minerals extraction which is required for renewable supply chains (Burke, 2020). Reflection on the proximity argument reminds us that the transition to a low-carbon future entails a restructuring that will impact both traditional fossil fuel communities as well as non-fossil fuel communities, and not simply employment structures (Le Billon and Kristoffersen, 2020). This has led some scholars to consider the importance of national or international responses to the transition, rather than focusing on affected communities (Pellegrini-Masini et al, 2020). Ethnicity and race have been raised also in the literature around their disproportionally affected minority groups from the air, in similar ways found in environmental justice (Adelman, 2013) and water pollution (Kayir, 2017) of long-term fossil fuel industries. Such studies have considered both direct and indirect impacts from, for example, workers in the industry to those living near their infrastructures.

Distributional justice goes beyond proximity. It also involves considerations of new spaces of vulnerability, recognition, risk and responsibility. All these dimensions are equally applicable to consideration within the context of just transition and its interpretation of the transition away from fossil fuels. The central jobs focus on just transition sheds light on the vulnerability of both young and old workers to major structural shifts in the job market (Dominish et al, 2019). These are most apparent in the post-COVID-19 pandemic recovery period (Cohen, 2020). In addition, gender-based recognition is critically needed for understanding the implications of the transition on under-recognised groups, as developed further in this volume in feminist justice theory (see Chapter 4). The indirect industries dependent upon fossil fuel industries are predominantly served by female employees (Wenham et al, 2020). Explicit consideration is required here. New responsibilities for communities to find post-carbon industries for employment have resulted in studies investigating the highly geographically variable rates of success. The risk of increased variation in communities that are preparing for the transition and those that are not is to dramatically increase in the coming decade (Steffen et al, 2020).

Procedural justice, or the requirement to consider due process as expressed in more legal accounts of justice, is a key consideration for just transition scholars. Most work focuses on enhancing the role of trade union representation as a potential solution to increased grievances among those affected by the transition away from fossil fuel industries (Newell and Mulvaney, 2013; Altintzis and Busser, 2014). I note here a critique raised in Chapter 1 on liberal justice on the predominance of such forms of justice linked to White, property-owning males often associated with fossil fuel or indeed renewable jobs. New forms of worker representation in an increasingly fragmented and underdeveloped renewable sector are of concern (Doorey, 2017). The focus here is not simply on industrial relations. It is increasingly about ensuring community-wide buy-in for major changes. One such example exists in the Netherlands where Groningen and its surrounding areas have been dominated by the natural gas industry. A series of commissions have been set up to engage those directly involved in the gas industry as well as others that depends on the industry locally to consult and engage in the new future. This led directly to local citizen assemblies that took place in 2022 and continue throughout 2023. Finding innovative ways to engage communities in employment and unemployment impacts of the transition is required in future research on just transition.

Conclusion

Justice scholarship does not sufficiently reflect upon the issue of transition. Sustainability appears to be understated in justice considerations, often resigned to intergenerational or environmental justice. Albeit not a central point of reflection for this book, it should be stated that sustainability is crucial for driving our attention towards critically reflecting on what we mean by transitions. Indeed, just transition scholarship itself sometimes does not explicitly reflect on this. But it offers a space to consider what is meant by the transition. A starting point provided by just transition is the consideration of sustainable outcomes and processes. Just transition's own development suggests that focusing on the transition away from fossil fuels and its associated journey towards renewables and low carbon infrastructures may be a fruitful point of departure.

Restoration is normally reserved for environmental-based considerations in applied justice research. It has been a central focus for environmental justice scholarship, especially with regard to post-mining clean-up activities as well as water and air pollution from polluting industries. The just transition concept brings a new dimension to the restorative justice argument. Its primary focus on workers' rights provides a useful way of considering how the loss of jobs, both directly and indirectly, for a community can lead to different types of inequality. This area of scholarship intimately connects spaces of restoration with that of transition. The journey towards a stated sustainable outcome involves necessary shifts in employment activities, structures and processes that are worthy of consideration within the inequalities framework. The just transition approach emphasises a workers' rights dimension of restoration beyond that of any other justice scholarship.

References

Abraham, J. (2017) Just transitions for the miners: Labor environmentalism in the Ruhr and Appalachian coalfields. *New Political Science*, 39(2), 218–231.

Adelman, D.E. (2013) The collective origins of toxic air pollution: Implications for greenhouse gas trading and toxic hotspots. *Indiana Law Journal*, 88(1), 273–337.

Agyeman, J. (2013) *Introducing Just Sustainabilities: Policy, Planning, and Practice.* London: Zed Books.

Altintzis, G. and Busser, E. (2014) The lessons from trade agreements for just transition policies. *International Journal of Labour Research*, 6(2), 269–294.

Brundtland, G.H. (1987) *Our Common Future: Report of the World Commission on Environment and Development.* Geneva, UN-Dokument A/42/427.

Burke, M.J. (2020) Energy-sufficiency for a just transition: A systematic review. *Energies*, 13(10), 2444.

Calvert, K., Greer, K. and Maddison-MacFadyen, M. (2019) Theorizing energy landscapes for energy transition management: Insights from a socioecological history of energy transitions in Bermuda. *Geoforum*, 102, 191–201.

Cha, J.M., Wander, M. and Pastor, M. (2020) Environmental justice, just transition, and a low-carbon future for California. *Environmental Law Reporter*, 50, 10216.

CleanEnergyWire (2020) *Spelling Out the Coal Exit – Germany's Phase-Out Plan.* https://www.cleanenergywire.org/factsheets/spelling-out-coal-phase-out-germanys-exit-law-draft

Cohen, M.J. (2020) Does the COVID-19 outbreak mark the onset of a sustainable consumption transition? *Sustainability: Science, Practice and Policy*, 16(1), 1–3.

Crowe, J.A. and Li, R. (2020) Is the just transition socially accepted? Energy history, place, and support for coal and solar in Illinois, Texas, and Vermont. *Energy Research & Social Science*, 59, 101309.

Dominish, E., Briggs, C., Teske, S. and Mey, F. (2019) Just transition: Employment projections for the 2.0C and 1.5C scenarios. In S. Teske (ed) *Achieving the Paris Climate Agreement Goals.* Sydney, Australia: Springer, pp 413–435.

Doorey, D.J. (2017) Just transitions law: Putting labour law to work on climate change. *Journal of Environmental Law & Practice*, 30(2), 201–239.

E2 (2020) *Clean Jobs America 2020: Repowering America's Economy in the Wake of COVID-19.* https://e2.org/wp-content/uploads/2020/04/E2-Clean-Jobs-America-2020.pdf

European Commission (2020a) *Green Deal: Coal and Other Carbon-Intensive Regions and the Commission Launch the European Just Transition Platform.* https://ec.europa.eu/commission/presscorner/detail/en/IP_20_1201

European Commission (2020b) *Recovery Plan for Europe.* https://ec.europa.eu/info/strategy/recovery-plan-europe_en

European Parliament (EP) (2020) *Just Transition Fund: Helping EU Regions Adapt to Green Economy.* https://www.europarl.europa.eu/news/en/headlines/economy/20200903STO86310/just-transition-fund-help-eu-regions-adapt-to-green-economy

European Trade Union Confederation (ETUC) (2018) *ETUC Resolution ahead of the Katowice Climate Conference.*

Evans, G. (2007) A just transition from coal to renewable energy in the Hunter Valley of New South Wales, Australia. *International Journal of Environment, Workplace and Employment*, 3(3–4), 175–194.

Evans, G. and Phelan, L. (2016) Transition to a post-carbon society: Linking environmental justice and just transition discourses. *Energy Policy*, 99, 329–339.

Friends of the Earth (FoE) (2017) *Just Transition: Is a Just Transition to a Low-Carbon Economy Possible Within Safe Global Carbon Limits?* https://friendsoftheearth.uk/sites/default/files/downloads/just_transition.pdf

Fuller, S. and McCauley, D. (2016) Framing energy justice: Perspectives from activism and advocacy. *Energy Research & Social Science*, 11, 1–8.

G7 (2018) *The Charlevoix G7 Summit Communique*. https://g7.gc.ca/en/official-documents/charlevoix-g7-summit-communique/

Gibbs, M. (2009) Using restorative justice to resolve historical injustices of Indigenous peoples. *Contemporary Justice Review*, 12(1), 45–57.

Goddard, G. and Farrely, M. (2018). Just transition management: Balancing just outcomes with just processes in Australian renewable energy transitions. *Applied Energy*, 225, 110–123.

Goodstein, J. and Butterfield, K.D. (2010) Extending the horizon of business ethics: Restorative justice and the aftermath of unethical behavior. *Business Ethics Quarterly*, 20(3), 453–480.

Gov (2018) *Climate Change Plan: Third Report on Proposals and Policies 2018–2023*. https://www.gov.scot/publications/scottish-governments-climate-change-plan-third-report-proposals-policies-2018/

Hall, N., Lacey, J., Carr-Cornish, S. and Dowd, A.M. (2015) Social licence to operate: Understanding how a concept has been translated into practice in energy industries. *Journal of Cleaner Production*, 86, 301–310.

Harrahill, K. and Douglas, O. (2019) Framework development for 'just transition' in coal producing jurisdictions. *Energy Policy*, 134, 110990.

Healy, N. and Barry, J. (2017) Politicizing energy justice and energy system transitions: Fossil fuel divestment and a 'just transition'. *Energy Policy*, 108, 451–459.

Heffron, R.J. and McCauley, D. (2018) What is the 'just transition'?, *Geoforum*, 88, 74–77.

Heffron, R.J. and McCauley, D. (2019) Beyond energy justice: Towards a just transition. In J. Jaria-Manzano and S. Borràs (eds), *Research Handbook on Global Climate Constitutionalism*. Cheltenham, UK: Edward Elgar, pp 23–31.

Howden-Chapman, P., Crane, J., Matheson, A., Viggers, H., Cunningham, M., Blakely, T. et al (2005) Retrofitting houses with insulation to reduce health inequalities: Aims and methods of a clustered, randomised community-based trial. *Social Science & Medicine*, 61(12), 2600–2610.

International Labour Organization (ILO) (2016) *Guidelines for a Just Transition towards Environmentally Sustainable Economies and Societies for All*. https://www.ilo.org/wcmsp5/groups/public/---ed_emp/---emp_ent/documents/publication/wcms_432859.pdf

Jenkins, K., Sovacool, B. and McCauley, D. (2018) Humanizing sociotechnical transitions through energy justice: An ethical framework for global transformative change. *Energy Policy*, 117, 66–74.

Jijelava, D. and Vanclay, F. (2017) Legitimacy, credibility and trust as the key components of a social licence to operate: An analysis of BP's projects in Georgia. *Journal of Cleaner Production*, 140(Part 3), 1077–1086.

Kayir, Ö. (2017) Violations of water rights, socio-ecological destruction and injustice in Turkey by hydro-electric power plants. *Transactions on Ecology and the Environment*, 200, 147–158.

Kenfack, C.E. (2019) Just transition at the intersection of labour and climate justice movements: Lessons from the Portuguese Climate Jobs Campaign. *Global Labour Journal*, 10(3), 224–239.

Lawrence, A. (2020) Conclusion: Just an energy transition – or a just transition? In *South Africa's Energy Transition*. New York: Springer, pp 115–152.

Lawrence, R. and Åhrén, M. (2016) Mining as colonisation: The need for restorative justice and restitution of traditional Sami lands. In L. Head, K. Saltzman, G. Setten and M. Stenseke (eds), *Nature, Temporality and Environmental Management: Scandinavian and Australian Perspectives on Peoples and Landscapes*. London: Routledge, pp 149–166.

Le Billon, P. and Kristoffersen, B. (2020) Just cuts for fossil fuels? Supply-side carbon constraints and energy transition. *Environment and Planning A: Economy and Space*, 52(6), 1072–1092.

Leijten, I. (2019) Human rights v. insufficient climate action: The Urgenda case. *Netherlands Quarterly of Human Rights*, 37(2), 112–118.

Mayer, A. (2018) A just transition for coal miners? Community identity and support from local policy actors. *Environmental Innovation and Societal Transitions*, 28, 1–13.

McAlinden, A.-M. (2011) 'Transforming justice': Challenges for restorative justice in an era of punishment-based corrections. *Contemporary Justice Review*, 14(4), 383–406.

McCauley, D. (2018) *Energy Justice: Re-Balancing the Trilemma of Security, Poverty and Climate Change*. London: Palgrave.

McCauley, D. and Heffron, R. (2018) Just transition: Integrating climate, energy and environmental justice. *Energy Policy*, 119, 1–7.

Mullally, G. and Byrne, E. (2016) A tale of three transitions: A year in the life of electricity system transformation narratives in the Irish media. *Energy, Sustainability and Society*, 6(1), 125–129. https://doi.org/10.1186/s13705-015-0068-2

Newell, P. and Mulvaney, D. (2013) The political economy of the 'just transition'. *Geographical Journal*, 179(2), 132–140. https://doi.org/10.1111/geoj.12008

OECD (2020) *Making the Green Recovery Work for Jobs, Income and Growth*. http://www.oecd.org/coronavirus/policy-responses/making-the-green-recovery-work-for-jobs-income-and-growth-a505f3e7/#endnotea0z14

Patterson, J. and Smith, J. (2016) Environmental justice initiatives for community resilience: Ecovillages, just transitions, and human rights cities. In B. Caniglia, M. Vallee and B. Frank (eds), *Resilience, Environmental Justice and the City*, 1st edn. London: Routledge, pp 82–99.

Peerla, D. (1999) Beyond the staples economy: The case for clean production and just transition. *Capitalism, Nature, Socialism*, 10(2), 153–157.

Pellegrini-Masini, G., Pirni, A., Maran, S. and Klöckner, C.A. (2020) Delivering a timely and just energy transition: Which policy research priorities? *Environmental Policy and Governance*, 30(6), 293–305.

Pollin, R. and Callaci, B. (2019) The economics of just transition: A framework for supporting fossil fuel-dependent workers and communities in the United States. *Labor Studies Journal*, 44(2), 93–138.

Presidency (2018) *Solidarity and Just Transition – Silesia Declaration*. https://cop24.gov.pl/presidency/initiatives/just-transition-declaration/

Preston, B.J. (2011) The use of restorative justice for environmental crime. *Criminal Law Journal*, 35(3), 136–159.

Rueckert, Y. (2018) The Global Unions and global governance: Analysing the dialogue between the international trade union organizations and the international financial institutions. *Economic and Industrial Democracy*, 42(3). https://doi.org/10.1177/0143831X18805846

Sareen, S. and Haarstad, H. (2018) Bridging socio-technical and justice aspects of sustainable energy transitions. *Applied Energy*, 228, 624–632.

Schot, J., Kanger, L. and Verbong, G. (2016) The roles of users in shaping transitions to new energy systems. *Nature Energy*, 1(5), 16054. https://doi.org/10.1038/nenergy.2016.54

Scottish Government (2020) £62 million fund for energy sector. https://www.gov.scot/news/gbp-62-million-fund-for-energy-sector

Scrase, I. and Smith, A. (2009) The (non-)politics of managing low carbon socio-technical transitions. *Environmental Politics*, 18(5), 707–726.

Shen, W., Srivastava, S., Yang, L., Jain, K. and Schröder, P. (2020) Understanding the impacts of outdoor air pollution on social inequality: Advancing a just transition framework. *Local Environment*, 25(1), 1–17.

Smith, L. (2013) Geographies of environmental restoration: A human geography critique of restored nature. *Transactions of the Institute of British Geographers*, 38(2), 354–358.

Snyder, B.F. (2018) Vulnerability to decarbonization in hydrocarbon-intensive counties in the United States: A just transition to avoid post-industrial decay. *Energy Research & Social Science*, 42, 34–43.

Steffen, B., Egli, F., Pahle, M. and Schmidt, T. (2020) Navigating the clean energy transition in the COVID-19 crisis. *Joule*, 34, 45–52. https://doi.org/https://doi.org/10.1016/j.joule.2020.04.011

Stevis, D. and Felli, R. (2015). Global labour unions and just transition to a green economy. *International Environmental Agreements: Politics, Law & Economics*, 15(1), 29–43.

Swilling, M., Musango, J. and Wakeford, J. (2015) Developmental states and sustainability transitions: Prospects of a just transition in South Africa. *Journal of Environmental Policy and Planning*, 18(5), 650–672. https://doi.org/10.1080/1523908X.2015.1107716

United Nations (UN) (2020) *Paris Agreement*. Paris: United Nations. https://unfccc.int/files/essential_background/convention/application/pdf/english_paris_agreement.pdf

van Welie, M.J. and Romijn, H.A. (2018) NGOs fostering transitions towards sustainable urban sanitation in low-income countries: Insights from transition management and development studies. *Environmental Science and Policy*, 84, 250–260.

Vilches, A., Barrios Padura, Á. and Molina Huelva, M. (2017) Retrofitting of homes for people in fuel poverty: Approach based on household thermal comfort. *Energy Policy*, 100, 283–291.

Weller, S.A. (2019) Just transition? Strategic framing and the challenges facing coal dependent communities. *Environment and Planning C: Politics and Space*, 37(2), 298–316.

Wenham, C., Smith, J. and Morgan, R. (2020) COVID-19: The gendered impacts of the outbreak. *The Lancet*, 395(10227), 846–848.

World Wide Fund for Nature (WWF) (2020) MEPs' dirty deal lets fossil gas into climate fund. https://wwf.panda.org/projects/one_planet_cities/sustainable_mobility/?851191/MEPs-dirty-deal-lets-fossil-gas-into-climate-fund

Yang, S., Zhang, D., Fu, J., Fan, S. and Ji, Y. (2018) Market cultivation of electric vehicles in China: A survey based on consumer behavior. *Sustainability*, 10(11), 4056. http://dx.doi.org/10.3390/su10114056

Zadek, S. (2019) Financing a just transition. *Organization & Environment*, 32(1), 18–25.

Zhang, X. (2014) Reference-dependent electric vehicle production strategy considering subsidies and consumer trade-offs. *Energy Policy*, 67, 422–430.

Conclusion

Johanna Ohlsson, Stephen Przybylinski and Don Mitchell

As the chapters in this volume show, theories of justice are broad, diverse, sometimes mutually exclusive, other times mutually supportive, and intensely debated. Justice, as we stated at the outset, is an intuitively simple idea, and, for most, an intuitively attractive, normative goal. Nevertheless, it is also an enormously complex concept, rooted not only in different philosophical and normative traditions, which generate often significantly different conceptions of what might constitute justice, but also in strong, often starkly opposed, political orientations. There is no neutral theory of justice. Nor is justice a question susceptible to technical solutions. A single measure of the justness of some act, structure or institution can never be developed. That is not to say there are not certain sets of common orientations towards key forms (substance, procedure, distribution, retribution), aspects (subject, object, domain, circumstances, principles) and realms (temporality, scope or scale, locus of concern, source(s) of harm) of justice that can be discerned in the welter of approaches identified throughout this volume. The goal of this concluding chapter is to elaborate some of these common orientations, examine their differences, and suggest other points of convergence and divergence in the traditions of justice theorising the authors have examined.

Points of convergence and divergence among philosophical and normative approaches

Before turning to a direct comparison of forms and aspects of justice among the normative traditions and applied fields of justice theorising, it is worth first examining some general connections and distinctions among the normative and philosophical traditions, beginning with liberalism as the benchmark for justice theories.

Liberalism as touchstone

It is no exaggeration to say that *all* contemporary western theorising on justice – even the most radical – has been and is shaped by the liberal tradition, either by critically engaging and developing its principles or by directly analysing the central tenets of liberalism. With its strong focus on the sovereign individual, on rights, freedoms and liberty, on the state as the primary domain within which justice claims are pressed, and on just *distributions* and *procedures*, liberalism is the benchmark against which justice theory is measured.

That liberal values and, more specifically, principles of justice, might be violated in practice is not exactly the point for many liberal theorists of justice (Chapter 1). Rather, the point is whether and how these principles can be improved and made more just and which of them should be given the most weight in designing a 'basic structure' for a liberal society. That is to say, much liberal justice theory (especially through the work of John Rawls, the most influential 20th-century liberal philosopher of justice) is concerned with defining the 'starting conditions' for a just society, theorising a set of normative and institutional procedures for safeguarding those starting conditions, or, when reality does not yet match the ideal, theorising the sorts of practices, in particular practices related to the just distribution of 'basic goods', that might push institutions closer to the ideal and thus create a more just world in which free and equal individuals have the equal possibility of living as free and equal individuals.

The concepts and principles of liberal justice theories are reflected in many of the traditions discussed throughout the volume. Libertarian theorists of justice (Chapter 2), for instance, adhere to basic liberal notions of freedom and protection of liberties and, as such, adopt much of the liberal framework that has been established over the centuries. At the same time, mainstream or right-leaning libertarians suggest that more mainstream liberal theorists get the weighting of principles wrong and in the process promote a society that unjustly limits the freedom and sovereignty of the individual, confiscates justly acquired property and wealth, permits state overreach and thus oppression, and misunderstands the moral basis and desirability of inequality in modern, free, society. For mainstream libertarian theorists, then, justice inheres not in equality or equity (as it does for liberal theorists like Rawls) but rather in *just deserts* of a very specific kind: deserts accrue to those who earn them.

Cosmopolitan theories of justice (Chapter 3) are also rooted in liberalism, if of a largely Kantian sort. Here, the importance of liberal subjectivity is not denied, but rather expanded to include not only citizens but also strangers, not only humans but other living beings and environments, and not only current generations but past and (especially) future ones. It asserts that the (liberal) Westphalian nation states are not always the appropriate domain for justice claims, given how social, political and economic relations, as well as the movements of human and other beings, extend across borders. If liberalism's (and libertarianism's) primary focus is on the *self*, the liberal individual, cosmopolitanism demands a focus on the self's relation to the *other* and the need to 'continuously question how the conditions of our coexistence and cohabitation can be improved', as Skillington puts it (Chapter 3).

Even bodies of scholarship tending to be critical of mainstream, liberal justice theorising necessarily engage critically with, if not promote, liberal principles. For instance, some feminist theorists of justice (Chapter 4) are avowedly liberal, as with Susan Moller Okin, who is centrally concerned with showing how liberalism does not live up to its own sacred tenets (for example, by subsuming women and their interests into men and their interests and thus violating the tenet of individual sovereignty and subjectivity). For such liberal theorists, (including Martha Nussbaum as well as Okin), holding liberalism to its own promises would entail a radical transformation of the basic conditions of justice. Other feminist scholars, however, take a much more critical stance towards liberalism's potential, questioning among other things its theory of the individual subject (asserting that individuals-in-relation-to-groups are the more apt subjects of justice theorising). Feminist theories of justice accept liberalism's focus on *(re)distribution*, but only to the extent that it is tightly bound to *recognition* and *representation*. With the second of these terms, feminism, similarly to cosmopolitanism, turns its attention from the self to the other and thus promotes a social and relational approach as essential in any full theory and practice of justice. With both the second and the third terms, similarly to postcolonial and some Indigenous thinking, it turns its attention to the question of *who* can be a subject of justice and under what terms. Finally, feminism tends to address its attention not only to the liberal state, but also to the family and the institutions of social reproduction (at all scales) more generally as the domain for justice claims.

Similarly, the more 'radical' traditions of scholarship detailed in Part I, such as the utopian socialists, anarchists, and Marx and Engels themselves (see Chapter 5), were centrally concerned with the conditions of possibility for the full development of humans *as* humans, for liberty, and for a just distribution. These traditions differ, however, as to what the role of distribution is in relation to the basic structure of society. They understand distribution to be subsidiary to, and an effect of, the conditions and relations of production and the forms of oppression, exploitation and injustice that arise within *them*. Crucially, as these traditions developed in the 20th and 21st century, they often did so in direct dialogue – or confrontation – with liberal theories of justice (as in the work of David Harvey), but they have also sought to transcend liberalism by rethinking justice as inhering somewhere between solidarity and 'the right to justification' (as with Rainer Forst). Recent Marxian inspired theories of justice have been centrally concerned with the conditions or procedures for creating *both* equal respect (recognition) *and* solidarity (as Jürgen Habermas put it) together with a broad equality of social conditions and a just distribution of social goods. Opposite to libertarian theories of just desert (distributed according to 'earning'), these radical theories, and Marxism in particular, found their theories of justice in the 'needs principle' ('from each according to their abilities to each according to their needs').

Given its fundamental aim of decentring the west as the primary locus of knowledge production, and geopolitical power, and given its presupposition that the rise of western liberal democracies was inextricably tied to its colonising and imperialist oppressions, postcolonial thinking is perhaps the furthest removed from liberalism as a touchstone (Chapter 6). Yet central, foundational claims of liberalism are not absent here either. Gayatri Spivak's 'planetarity' is not far from Kantian cosmopolitanism; postcolonialism's insistence that the subaltern be heard and accounted for is not that far removed from Mary Wollstonecraft's (or Susan Moller Okin's) insistence that woman be afforded the status of full subjectivity. Theories of epistemic injustice (such as those outlined in Chapter 6, but which are also vitally important in feminist theories) presuppose, as does liberalism, a public sphere of rational discursive practice in which claims can be heard and adjudicated; postcolonialism largely imagines these to be decentred not eliminated.

To a large extent adopting much of the liberal paradigm on rights and liberties, Indigenous approaches to justice (Chapter 7) critically suggest ways to improve the liberal state, by calling for inclusion and representation, or autonomy and self-determination. Yet, at the same time, the material relations and actions of the liberal state towards Indigenous peoples naturally underscores a full rejection of the liberal principles mentioned here. As such, some Indigenous approaches entirely dismiss the dominate discourse of liberalism, instead promoting ontological practices of justice that resonate with epistemological understandings of what constitutes just relations within and outside of Indigenous communities, particularly as seen through relations of environmental stewardship. Like some of the radical approaches, Indigenous scholarship on justice is critical of the strong individualistic focus in liberalism, stressing instead more relational approaches towards healing and reconciling social and ecological injustices. What is specific for Indigenous approaches is the combination of rights and the broader understanding of human nature also including the environment in an intergenerational way, opening up for nature having rights.

Lacking a comprehensive theory of justice of its own, the Capabilities Approach (Chapter 8) generally (if often implicitly) adopts a liberal, or liberal cosmopolitan (as with Nussbaum), foundation and provides a methodology for realising it in 'underdeveloped' societies. As in liberalism, the Capabilities Approach is centred on individuals, their well-being and ability to make rational decisions about the shaping of their lives, and the question of what might hinder or enable these abilities.

Liberalism has thus defined many of the axes around which justice theories of all stripes, however ontologically, normatively and politically oriented, turn. However, liberalism has hardly defined the full scope of justice theorising that even the rather limited and selective analyses of the chapters make clear. For us, it is important to recognise both the common axes *and* the broader scope of impact liberal theorising has made. Without understanding these common axes, theories of justice risk losing track of a central point of justice, which is the full, equitable development of humans in and among their societies and environments. At the same time, without understanding the full scope of justice theorising, beyond those rooted within liberal principles, theories of justice risk becoming victims of their own ideologies, missing their silences, blind spots and potentially oppressive tendencies.

An ideal state of justice or actually-existing conditions of injustice: justice orientations

Traditions of justice theorising differ in whether their focus is on the ideals of justice or the facts of injustice and therefore in whether they begin from seeking to determine an ideal state of justice or undertaking a materialist analysis of 'actually-existing conditions'. Typically, these are differences of degree, emphasis or starting point; they are hardly mutually exclusive.

The mainstream of liberal theorising about justice has been centrally concerned with understanding the condition of justice: what arrangement of institutions and relations will promote, for example, a just distribution of goods, life chances, offices, and so on. It is less directly concerned with theorising *injustice*. More radical variants of liberalism, like (right-wing) libertarianism, even sometimes argue that what many liberals might decry as unjust – like gross inequality – is in fact just. Inequality, for them, is not only natural, but inequalities of the right kind (for example, of income) are a marker of a just society. Full equality would be a mark of an unjust society that inhibits human liberties. But in both, the preoccupation of theory is with identifying the ideal constituents of a just society, as it also is (though from a very different orientation) with much cosmopolitan thinking.

By contrast, many feminist, Marxian and other radical theories are typically more materialist in orientation and often begin by identifying conditions of 'actually-existing' injustice. Perhaps this orientation is best expressed by the (in fact liberal, but still radical) feminist Susan Moller Okin who insisted that 'a theory of justice must concern itself not with abstractions or ideals of institutions but with their realities' (Moller Okin, 1989; Chapter 4). By partial contrast, postcolonial, anarchist and utopian socialist brands of radical theory tend more to be idealist in orientation (than, for example, Marxian approaches) with some, as with Spivak, trading in a strong brand of neo-Kantian idealism, and others, like Fourier, seeking to imagine the ideal conditions for a society that transcends injustice.

The Capabilities Approach sits close to these more radical approaches to justice in that it sees itself as non-idealist, but it differs from them in refusing to specify in any normative way *which* 'realities' need to be addressed (Nussbaum's work is an exception in this regard). For the Capabilities Approach, these realities do not *necessarily* include patriarchy and gender oppression (as in feminism), the conditions and relations of production and exploitation (as in Marxism), or the ongoing legacies of colonialism, for example, epistemic injustice (as in postcolonialism and Indigenous approaches). They only include what locally embedded people determine they ought to include.

Table 1 gives some sense of how the major traditions of justice theorising we have surveyed can be arrayed in relation to an ideal state of justice or actually-existing conditions of injustice.

As is visualised in the table, most of the traditions examined in the chapters of Part I are centred on ideal state of justice, or more explicit normative reasoning. Important to notice is that these are the dominant themes in each tradition and there are examples of traditions centring on both justice and injustice at the same time. The traditions that more clearly centre on actually-existing conditions and have a focus on injustice (the Capabilities Approach, feminism and Marxism) tend to vary in which way they relate to and critique liberalism. Again, these are not clear-cut but rather a way of organising the dominant trends in the traditions explored in Part I. For instance, Indigenous approaches are often deeply embedded in the actually-existing conditions for injustice and in how they are connected to the material earth as well as resources and the human–nature connection, but many of

Table 1: Justice orientations

	Ideal state of justice	Actually-existing conditions
Starting focus on justice	Liberalism Libertarianism Cosmopolitanism	
Starting focus on injustice	Anarchism Utopian socialism Postcolonialism Indigenous approaches	The Capabilities approach Feminism Marxism

the Indigenous scholars working on justice theory simultaneously have a clear focus on the normative reasoning of how this should be accomplished.

The pairs represented here are not mutually exclusive but they do orient attention in different directions and lead both to different normative traditions (each with their own internal consistency) and to potentially different modes of intervention. Understanding the nature and importance of these orientations, both in how they shape the internal consistency of single justice traditions and how they indicate differences among these traditions, has implications not only for assessing current and ongoing development initiatives or policy proposals, but also for assessing any sort of intervention into current conditions and the planning of new initiatives.

New policy proposals can often be developed within a somewhat delimited field, and the current social and environmental state of being seems to pay particular attention to issues of the environment, climate, energy systems, geographical space, landscape, intergenerational temporality, as well as a green just transition. Such societal relevance adds to the reasons for Part II being centred on these fields. It is now time to turn to a synthesising discussion of the applied fields of justice.

Points of convergence and divergence among the applied fields

An identifiable field

After having discussed a few overarching themes across the philosophical and normative approaches to justice, we turn in this section to discuss a few points of convergence and divergence among what we call the 'applied fields', the topics for the chapter in Part II. One overarching aspect connecting these sets of discourses is that they all focus on an identifiable *field* at the interface of human and natural relations. For instance:

- Environmental justice is concerned with power-laden *social relationships as mediated through environmental harms*. It focuses on the disproportionate distribution of life-limiting environmental harms and the racist and marginalising practices that promote this disproportionality.
- Climate justice is concerned with the unequal effects of, and the power-laden processes producing, *climate change*. It seeks to understand how to apportion responsibility, to account for future generations, and to promote and protect the interests of those most affected by climate change's adverse effects.
- Energy justice is concerned with the overall *energy system* and understands energy as a particular good that is instrumental to modern society. Many theorists argue it is essential to human flourishing. Energy justice research thus seeks to understand the forces that shape the production, transmission,

use and waste of energy, and how these may disproportionately advantage or disadvantage particular groups, communities, and so forth.

- Spatial justice is concerned with *geographical space*. It focuses on the social relations that create geographical space, how geographical space shapes social relations, and the ways that matters of justice and injustice are necessarily entangled within this.
- Landscape justice is concerned with humanly transformed nature: the *landscape*. It understands landscapes as composed of both shape and structure (morphology) and representation (ideologies, ideas, senses of belonging). It inquires after what would constitute just social relations in shaping, preserving and living in landscapes.
- Intergenerational justice centres on *temporality* across generations, both historical and future beings. The focus is centred on relations across previous, current and future generations, while tackling societal and environmental challenges.
- Just transition is concerned with the *transitions* from an unsustainable and oftentimes unjust way of organising society, especially through the lens of labour, in order to bring about more sustainable futures for organising industries and societies.

These fields are not mutually exclusive; they sometimes overlap in significant ways. Nonetheless, research in each field has often, though not always, developed in relative isolation. This in itself indicates the potential for new research. For instance, which new insights are generated when analysing the central issues discussed in energy justice from a feminist perspective? Or, are aspects of landscape justice possible to analyse through a libertarian lens, and if so, how? While there are references to energy systems in landscape justice research, and while climate justice is also a matter of environmental justice, at the level of theoretical or conceptual development, each area has a relatively autonomous genealogy. Yet there are also intersections or convergences, among their divergent aims and foci.

As indicated in the Introduction, each of these applied fields operates in relation to a set of *realms* defined by:

- the *temporalities* they work in;
- the *scope or scale* of the phenomena they examine;
- their specific *locus of concern*;
- the *source of the harms* they are concerned with.

In common with the theories of justice examined in Part I, the applied fields also offer insights into the *forms* of justice, as we discussed in the Introduction. To reiterate the point we made earlier, these categories offer a way of structuring a justice analysis and represent the main forms of justice discussed in contemporary justice scholarship. The forms are:

- *substantive* justice;
- *procedural* justice;
- *distributive* justice;
- *retributive* justice;
- *recognitional* justice.

Here, we want to emphasize *recognitional justice* as a central form of justice. This is an already established form in several of the justice traditions; while it is still not present in others. What we have noticed while working on this volume is that recognition is one of the key aspects for implementing several of the other forms of justice, which also justifies an increased focus.

Next to the forms of justice, we make use of the *aspects* of justice also in Part II:

- the *subject* of justice (who);
- the *object* of justice (what);
- the *domain* of justice (where);
- the *social circumstances* of justice (when); and
- the *principles* of justice (how).

While discussion of these realms, forms and aspects of justice are sometimes implicit in the literatures surveyed in each chapter, they are nonetheless discernible. And when they are not – when a field has not engaged with a particular realm (for example, temporality) or form (for example, retribution) – that absence is important to understand too, as it offers an opportunity for researchers to speculate as to, first, why that might be, and second, to explore how the applied field might be developed if it were to focus more closely on that particular facet of justice discourse and practice.

For our purposes now, however, the identification of if and how each applied field has addressed the realms, forms and aspects of justice allows for an examination of the points of convergence, divergence and thus synergy among them.

Topical or methodological focus of the applied fields

What connects these applied fields is their understanding of justice and injustice as being not simply environmental, climatic or natural, but as necessarily social as well. Conversely, justice and injustice are not *only* social, but are also environmental, climatic or natural, even while in some (for example, anthropocentric environmental justice) the social is given ontological priority, while in others nature is (ecocentric environmental justice). To put this another way, in these fields, justice is not simply about restoring problematic environmental conditions or reducing climate emissions in the abstract, it is also about recognising and ameliorating the social inequalities that were already present and that these 'natural' conditions exacerbate.

Topically, each field is concerned with the interface, or rather a range of interfaces, between human and natural worlds (though spatial and intergenerational justice and the just transition are partial exceptions here), with the nature of social relations that shape and is shaped by that interface, and how together these interfaces and social relations structure the conditions of possibility for justice and injustice. Each of the applied fields also offers valuable methodological orientations that ought to be synergistic:

- Environmental justice asks researchers to be attentive to how socially unjust relations are mediated through environmental harms, while those harms themselves are unavoidably social in origin and character.
- Climate justice turns attention towards future generations in relation to past and present generations and thus towards the complexities of intertemporality; it also focuses attention on the mismatch between the scales at which climate changing practices are produced, governed and have their effects.
- Energy justice concerns the right to energy and the right to be free from harms associated with the production, transmission and use of energy. It trains attention on systems, rather than parts in isolation.
- Spatial justice demands a close focus on the complexity of geographical space (the interdigitation of its absolute, relative and relational aspects) together with geographical scale since both of these define the field of action within which socio-environmental processes play out (household, neighbourhood, city, region, nation state, globe).
- Landscape justice pays particular attention to the production of place and territory, the labour processes that are central to it, and the procedures through which landscapes are maintained and transformed. It further demands a close attention to the affective meanings that landscapes have in communities' lives.
- Intergenerational justice is, similar to climate justice, centred on intertemporality and its material as well as potential effects on past, current and future societies and environments.
- Just transition focuses on the processes of transition from unsustainable practices of industrialisation to more sustainable ways of creating growth and job markets. By initially paying particular attention to labour

conditions, the emerging field of just transition is widening its scope towards societies and environments at large.

We address these different methodological foci here under convergences and synergies not because they necessarily *have* converged or been synthesised in the literature, but because there are potentially huge benefits – including especially for understanding the relative justness of various initiatives, actions or structures – when and if they do converge and are synthesised.

The common focus on an identifiable field in each of these applied theories (environment–human interactions, climate change, energy systems, space, landscape, temporalities and transitions) together offer a reasonably comprehensive coverage of the crucial variables at stake.[1]

Finally, propositions arising in each of these fields have to a greater or lesser extent begun to be institutionalised within state and global governing bodies and non-governmental organisations. For example, discourses and practices of environmental justice have been highly institutionalised in some nation states, principles associated with climate and intergenerational justice are increasing concerns among national and international governing bodies, landscape justice finds partial institutional expression in the European Landscape Convention, some states are beginning to adopt energy justice policies, and spatial justice is reflected to a degree in various Right to the City ordinances and policies (for example in Brazil's constitution and in Mexico City's City Charter). Just transition ideas have showed the ability of non-governmental organisations to function as global drivers for change, when states and state-based organisations have been slow.

All applied fields also have been shaped by, and in many cases continue to respond to, social movements or activism: environmental justice (born as a civil rights endeavour), climate justice (though movements too numerous to list), energy justice (through campaigns against energy poverty), spatial justice (linked to Right to the City movements), intergenerational justice (lead theme for many youth groups), just transition (developed in and though the international labour movement) and, though to a lesser extent so far, landscape justice (in relation to campaigns seeking to assure national and regional policy aligns with the prescriptions in the European Landscape Convention). These linkages are important because they sometimes directly shape research frontiers, but also because they are a primary way in which the theories of justice are and remain 'applied'.

The potential that was just named is not necessarily easily achieved, however. There are, after all, significant ontological and epistemological differences among, and within, each of these seven applied fields (Chapters 9–15). Environmental justice and climate justice are not the same, for example. Sometimes the divergence between the temporal scale appropriate for addressing environmental injustices (the human lifetime, or perhaps two generations) and the interventions that focusing on such a time scale implies, will clash directly with the temporal scale appropriate for assessing and addressing climate change injustices (many generations in the future). A focus on the systemic aspects of energy systems might run head on into a focus on the localised participatory prerogatives associated with efforts to enhance landscape justice. In addition, assertion of a 'right to landscape' may entrench, rather than ameliorate, the racist politics of privilege that some contemporary environmental justice research warns against.

In other words, careful attention to where these discourses contradict each other is as important as where they are potentially mutually beneficial. A concept like 'the right to landscape' cannot simply be *applied* to discourses of environmental or climate justice, but will necessarily need to be *transformed*, given the different ontologies and epistemologies that define the fields.

Therefore, it is helpful to note there are particular divergences among and within the fields that stand out as particularly important:

- Environmental justice's focus on systematic, environmental racism and other forms of oppression and marginalisation finds something of an echo in spatial and landscape justice discourses, but is considerably more developed there than in those or the other fields. While climate justice is centrally focused on questions of inequality, as well as on the legacies of colonialism (at least to some degree), it is less focused than is environmental justice on questions of how intersectional oppressions can be and have been deployed by powerful interests to naturalise injustice.
- Climate and intergenerational justice are the fields most concerned with the (distant) future, and its focus on the limits and problems associated with intertemporal discounting is not matched by the other fields.
- Landscape justice stands out from the others in its incorporation of aesthetic matters as well as in its inclusion of matters related to representation in the sense that a picture or photograph represents (rather than – or, rather in addition to – representation in the sense of political representation).
- Most of the fields we examined are concerned with collectives: marginalised peoples, communities, places, states, and so on. This is not to say that individuals (or the rights of individuals) are not important, but rather that such individuals are typically understood as part of and in relation to collectivities. This is particularly clear in the field of just transitions.
- The focus on *systems* sets energy justice apart from the other applied fields, and might even be anathema to some landscape justice theorists, for example, whose concerns lie not necessarily with systems but rather in understanding how 'moral orders' shape the ways people interact with place and society. Another point of divergence is with environmental justice, where discourses on environmental harm generated by the production, transportation and use of energy often draw comparisons

to environmental justice, sometimes to the point of calling energy justice redundant. Whereas the concepts might overlap in some respects, energy justice also concerns the (potential) role energy plays in human flourishing, a concern largely absent from the environmental justice literature.

There are other important divergences and distinctions as well. For example:

- Climate justice concerns human institutions and practices.
- Questions of scale are especially important in the fields of spatial justice, landscape justice and climate justice, and perhaps to some extent in environmental justice.
- Discourses of spatial and landscape justice have been developed by and indeed largely confined to research in a single or a small handful of closely related disciplines, while the discourses in climate, environmental and energy justice are much more interdisciplinary and thus often are more broadly and deeply developed.

As with the convergences discussed previously, these divergences should be understood as opportunities for exploring synergies, but to also indicate this may not be easy. What the chapters in Part II showed us is that justice theorising/analysis cannot be thought of as some kind of a restaurant menu (pick one ontology from column A, two from column B, and one from column C). That is not how theoretical synthesis or political practice works. Rather ontological (and epistemological) incompatibilities have to be logically worked through. A theory of justice, related to the environment, landscape or climate, cannot be libertarian and Marxist at once; it cannot be postcolonial and liberal. Neither can it easily be ecocentric and also have as its central focus point anthropocentric concerns like the elimination of environmental racism or biocentric and work with climate justice theories developed in relation to current discourses of the Anthropocene.

Returning to the forms and aspects of justice

As we stated in the Introduction to the volume, and already in this concluding discussion, there are a range of ways in which the forms (substance, procedure, distribution, retribution, recognition) and aspects (subject, object, domain, circumstances, principles) of justice are understood in the different normative approaches and applied fields. We have deliberately not applied the realms of justice to the normative traditions in Part I, as they have been developed inductively throughout our work with the applied fields in Part II. Therefore, we will leave the realms for now, and return to them after summarising the forms and aspects.

269

Now we want to array the relationship among these approaches in a slightly different way, by bringing them together, emphasising the similarities and connections, rather than the differences among them. We do so by offering a series of tables to more neatly categorise their connections. As a result of this, much nuance and detail is missed. In exchange, however, some insight into points of convergence in justice theorising emerge. We see this as instructive and helpful for teaching situations, but also for an indication of gaps for future research. As we see it, the tables have pedagogical value just because of their simplification. For instance, this could generate fruitful discussions about what the tables do *not* say or convey.

Placement in each cell indicates a *predominant* orientation and is only indicative; placement also does not indicate if the relation is positive or negative (for example states are the *object* of some cosmopolitan theorising, but often because at the scale of the nation, such states might threaten cosmopolitan justice ideals). Most traditions spill across many of the cell walls, some and sometimes much of the time. (CA indicates the Capabilities Approach, PCA indicates postcolonial approaches, EJ indicates environmental justice, and JT indicates just transition.)

We use the aspects and forms to compare the foci of the approaches central to Part I and the applied fields central to Part II. For instance, we address questions such as: which fields focus on the individual as a subject of justice, and which focus on other species or nature as a whole? Which are oriented towards local domains and which towards global ones? The particular foci of each aspect of justice (for example, individuals or vulnerable peoples as subjects of justice) were also derived from our analyses of the schools of justice in Part I and one of the points here is to see whether and how those foci make sense for understanding the analyses found in the applied fields of justice.

Tables 2–11: Forms and aspects of justice

These tables are useful to scholars and students as a helpful quick reference. They allow for comparison both within and across aspects and forms of justice, and within and across the normative traditions and applied fields. For instance, if the concern is with questions of *distribution* in relation to *future generations*, or questions of *rectification* and *responsibility*, or how *rights* are or are not foundational to *justice*, the tables will allow researchers to quickly determine which sets of theories or justice tradition might provide helpful insights, and which – based on the current state-of-the-art – are unlikely to. As should be clear from the foregoing chapters, a reference guide like this can be no substitute for the careful consideration and analysis of the presuppositions, ontological and epistemological foundations, and the political orientations of the theories of justice and injustice developed within the applied fields. In addition, these tables suggest *potential* compatibilities or synergies among applied discourses, not necessarily actual ones.

Table 2: Substantive justice

Substantive justice	Fair distribution	Equality of opportunity	Individual sovereignty	Equality of persons	Just treatment of groups	Human capabilities	Needs principle	Freedom from oppression, just treatment of groups	Just mode of production	Focus on injustice	Rectification of past wrongs
Justice tradition	Liberal Cosmo Feminist	Liberal	Liberal Libertarian	Cosmo Feminist PCA Indigenous	Feminist PCA Indigenous	CA Indigenous	CA Marxist	Anarchist Socialist Marxist Feminist PCA Indigenous	Marxist	CA Feminist Marxist PCA Indigenous	PCA Indigenous
Applied field	Spatial EJ Energy Intergen	Energy JT		Spatial Landscape Climate EJ JT Intergen	Landscape EJ JT Intergen		EJ Energy JT	Landscape EJ	JT	Spatial Landscape Climate EJ Energy Intergen JT	

Table 3: Procedural justice

Procedural justice	Rights-based	Non-subsumption of interests	Equal treatment	Market	Development, needs based	Deliberative democracy/right to justification	Representation/recognition	Truth and justice process	Rectification
Justice tradition	Liberal Libertarian Cosmo Indigenous	Liberal Feminist Cosmo	Liberal Feminist Cosmo	Libertarian	CA	Marxist PCA	Feminist Marxist PCA	PCA	PCA
Applied field	Landscape Climate EJ Energy Intergen JT		Spatial Landscape Climate EJ Energy Intergen	Landscape Energy JT	Energy Intergen	Spatial Landscape Climate EJ Energy	Spatial Landscape Climate EJ Intergen		Intergen

Table 4: Distributive justice

Distributive justice	Difference principle	Just acquisition	Recognition	Ownership of means of production	Expertise/ knowledge	Rights/ legal resources	Well-being
Justice tradition	Liberal	Libertarian	Cosmo Feminist Marxist PCA Indigenous	Anarchist Marxist Indigenous	Cosmo PCA Marxist Indigenous	All	CA
Applied field	Spatial Energy Intergen		Landscape Climate EJ Energy Intergen JT		Spatial Landscape Climate EJ Energy	Spatial Landscape Climate EJ Energy Intergen	Climate Landscape EJ Energy Intergen

Table 5: Retributive justice

Retributive justice	Tribunals	Court challenges, prosecutions	Compensation schemes	Resettlement agreements	Responsibility/ reparations	Memorialisation
Justice tradition	Cosmo PCA	Liberal Libertarian Cosmo Indigenous	Liberal Libertarian Cosmo PCA Marxist Indigenous	Liberal Cosmo PCA	Feminist PCA Indigenous	PCA
Applied field		Climate Energy Intergen	Climate EJ Energy JT	Climate EJ Energy JT	Landscape Climate EJ Energy Intergen	

Table 6: Recognition justice

Recognition	Epistemic	Rights	Decision-making/participation	Responsibility	Respect for the subject of justice
Justice tradition	Indigenous Feminist PCA	Liberal Libertarian Cosmo Feminist Marxist PCA Indigenous	Liberal Cosmo Feminist Marxist PCA Indigenous	Liberal Cosmo Feminist Marxist PCA Indigenous CA	Liberal Cosmo Feminist PCA Indigenous
Applied field	Climate EJ	EJ Climate Spatial Landscape Intergen Just transition	Climate EJ Energy Landscape Intergen	Climate EJ Energy Intergen	Climate EJ Energy

Table 7: Subjects of justice (who)

Subject of justice	Individual(s)	Individuals: in-groups, classes, families, community	Humanity in general	The other, strangers, subalterns, exploited and oppressed peoples, the vulnerable	Other species	Nature	Future generations
Justice tradition	Liberal Libertarian CA Cosmo	Feminist Marxist Socialist Anarchist	Feminist Cosmo Marxist	Feminist Cosmo Marxist PCA	Cosmo PCA	Cosmo PCA Anarchist	Cosmo
Applied field	Climate EJ Energy	Spatial Landscape Climate EJ Energy	Spatial Climate	Spatial Climate EJ Energy	Climate	Landscape	Climate EJ Energy

Table 8: Objects of justice (what)

Object of justice	The market	Basic structure	Families	States	Movement of change	Harms	Governance institutions	Production/ reproduction	Relations of power
Justice tradition	Libertarian	Liberal Feminist	Feminist	Liberal Libertarian Cosmo PCA			CA Cosmo	Feminist Marxist Socialist Anarchist	Marxist PCA
Applied field					Spatial Climate EJ Energy	Spatial EJ Climate Energy	Landscape Climate EJ Energy	Spatial Climate Landscape Energy	Spatial Landscape Climate EJ Energy

Table 9: Domain of justice (where)

Domain of justice	The market	Global system	Supra-state	Nation-state	Local communities	Peripheries
Justice tradition	Libertarian	Cosmo Marxist PCA	Liberal Cosmo Marxist Socialist	Liberal Libertarian Cosmo Marxist Socialist	Anarchist CA Cosmo Feminist	PCA
Applied field		Climate Energy	Climate	Landscape Climate EJ Energy	Spatial Landscape Climate EJ Energy	

Table 10: Social circumstances of justice (when)

Social circumstances of justice	Constitution of society	Moments of exchange/ distribution	Moments of production/ reproduction/ development	Moments of encounter with difference/ stranger/other	Moments of recognition/ representation/ memorialisation	Moments of rectification/ responsibility	Development initiatives
Justice tradition	Liberal Anarchist Socialist	Feminist Libertarian Liberal Marxist	Feminist Marxist	Feminist Cosmo	Feminist Marxist PCA	Feminist PCA	CA
Applied field	Spatial Landscape Climate EJ Energy	Landscape EJ Energy	Spatial Landscape Climate EJ		Climate	Landscape Climate EJ Energy	Energy JT

Table 11: Principles of justice (how)

Principles of justice	Difference principle	Minimalist state	Hospitality	Universality/ pluriversality	Equality (within difference), solidarity	Responsibility/ rectification	Recognition/ representation	Transformation of means of production	Liberation
Justice tradition	Liberal	Libertarian	Cosmo	Cosmo	Cosmo Feminist Marxist Anarchist	Feminist PCA Indigenous	Feminist Cosmo PCA Indigenous	Marxist Anarchist Socialist	Marxist Feminist PCA Indigenous
Applied field	Energy JT		Climate	Climate	Spatial Landscape EJ Energy	Spatial Landscape EJ Climate	Climate EJ JT	EJ Climate JT	EJ

We encourage readers to make use of the discrepancies explored in the chapters and visualised here in the tables for developing new research. One of the areas for future research we encourage further engagement with is more critically exploring how the ideals and arguments emanating from the normative traditions of justice are utilised within the applied fields. For, as these chapters show, the normative traditions offer excellent foundations for discussing the content of justice in theory, but the reality of situations as examined empirically may leave much to be desired conceptually. Several of the applied fields are still emerging, contributing new concepts to help frame justice analysis, and thus promote new research areas. Again, there is no correct approach but some cross-fertilisation seems promising.

The synthesis of forms and aspects with the realms of justice further guide research on justice and injustice. We now turn to a discussion of the realms of justice and to visualise their role in the applied themes of justice research.

Considering the realms of justice

The tables and discussion have already indicated the range of ways in which the forms (substance, procedure, distribution, retribution, recognition) and aspects (subject, object, domain, circumstances, principles) of justice are understood in the different normative (Part I) and applied (Part II) theories we have investigated. Whether the applied fields directly appeal to normative theories of justice or not, they all to various degrees address a set of themes, or what we call 'realms', around and through which theories of justice (and not just injustice) are being or could be constructed. Thus, as first mentioned in the Introduction, a final way in which we have found useful for categorising or framing justice approaches is through the aforementioned realms: *temporality*, *scale/scope*, *locus of concern*, and *sources of harm*.

By adding another conceptual layer, the four realms of justice, we suggest a way of structuring comparison between the applied fields. We also summarise the connections and divergences among them in a series of tables. As in earlier tables, much nuance and detail are missed, and placement in each cell indicates a predominant orientation and is only indicative; again, placement also does not indicate if a relation is positive or negative. Most traditions spill across many of the cell walls.

Tables 12–14: Realms of justice and the applied fields

For example, the first realm, time, or temporality, is to be of importance because the theories in the applied fields have slightly different temporal

Table 12: Temporality in the applied fields

Temporality	Historical focus	Contemporary focus	Future focus	Human time	Natural time	Long-term processual time	Immediacy
Applied field	Landscape EJ Energy Intergen JT	Spatial Landscape Climate EJ Energy JT	Climate EJ Energy Intergen JT	Spatial Landscape Climate EJ Energy Intergen	Climate Intergen	Landscape Climate	Spatial Landscape Climate EJ Energy JT

Note: We do not array the 'locus of concern' in these tables, because, in essence, the locus of concern in each of these fields is unique: it is what sets each apart from the others. EJ is environmental justice, and JT refers to just transition.

Table 13: Scale and scope of the applied fields

Scale/scope	Local	Urban	Regional	Nation state	Global
Applied field	Spatial Landscape EJ Energy JT	Spatial Climate EJ Energy	Landscape Climate EJ Energy	Landscape Climate EJ Energy JT	Spatial Climate Energy Intergen JT

Table 14: Source of harm in the applied fields

Source of harm	Pollutants	Production systems	Distribution systems	Consumption systems	Social structures (for example, states, institutions, corporations)	Social practices (for example, racism, economic exploitation)	Humanity in general
Applied field	Climate EJ Energy	Spatial Landscape Climate EJ Energy JT	Spatial Landscape Climate EJ Energy JT	Climate Energy JT	EJ Climate Energy Spatial Landscape Intergen JT	EJ Spatial Landscape Energy JT	Climate Intergen

foci. Most applied areas focus on the contemporary situation, while a clear future-oriented outlier is climate justice, and landscape justice incorporates a stronger historical orientation than the others. Intergenerational matters are perhaps strongest in (again) climate justice and most muted in spatial justice.

We have sought to capture at least some of the areas of convergence and divergence in relation to these four realms in Tables 12–14. In the tables, we re-array these findings to emphasise the commonalities rather than the differences among the schools. For each realm, we have identified a set of facets – for example, a range of possible temporalities – and arrayed the schools in relation to them.

We find this schematic overview helpful as it visualises some of the most central areas of focus within the applied fields. When cross-examining the fields as such, it becomes clear that the main focus for the different fields sometimes overlap and sometimes contradict one another. Spatial justice, for instance, pays little to no attention to the temporal relations of injustice, while for intergenerational justice, this is its *raison d'être*. What is important to remember, though, is that these divergences point to openings for researchers to bring these foci together in their justice analyses. For, there is not necessarily a contradiction between having an historical, contemporary and future oriented focus. What matters most is how these foci are actualised in to illustrate how justice may be understood and implemented in practice. We briefly discuss what that implies in the final section.

Justice in practice

As the chapters show, the seven applied fields selected (environmental justice, climate justice, energy justice, spatial justice, landscape justice, intergenerational justice and just transitions) and the realms of justice they explore offer ample, if always incomplete and sometimes flawed or contradictory, grounds for exploring what 'just' can mean *in practice*. For example, a project that appears 'just' in the terms established in the energy justice field, might conversely contain aspects that appear 'unjust' when examined in relation to the landscape and its dynamics, or, conversely, the environmental justness of a set of policies related to sustainable development might be strengthened when supported by policies seeking to enhance climate justice.

In other words, these seven fields offer a *methodology* or approach for understanding and examining justness, if a quite complex one. However, two significant cautions are in order: first, as we have stressed in a number of places, these applied justice fields are themselves often incomplete, especially in relation to a solid, ontological grounding in moral and political philosophy of justice. Without such a grounding, they can be subject to problematic manipulation and can become more susceptible to the raw exercise of power. 'Between equal rights, force decides', as Marx put it, and the same axiom is true for theories of justice unless the hard work is done to show why one or another approach to justice is more justifiable given both its ontological presuppositions and the context in which it is being applied.

Second, there are other applied fields beyond those surveyed here. We have deliberately focused on the most prominent fields of justice scholarship, but emerging fields such as mobility or transportation justice, ecosystem or food justice could also be of great interest. This, we see as yet another sign of the continuous development within justice scholarship.

Conclusion

Summing up a volume such as this is no easy task. It would be an impossible and fruitless endeavour to point out all potential connections and departures between the various approaches detailed throughout this volume. There are simply too many terms, concepts and approaches to justice theorising to effectively capture any one true essence. Further, we do not endorse any one position surveyed herein over another and do not suggest one approach better identifies and assesses matters of injustice more effectively. Instead, we argue that the chapters included in this volume are best seen as starting points for further discussion, not as complete or comprehensive accounts of justice theory. With that being said, we think that between the chapters and our concluding discussion bringing them briefly into conversation, we have created a solid foundation for understanding the multifaceted concept of justice and its connection to different traditions in normative and applied fields.

We hope that by highlighting these 'ways of seeing' that we have prompted readers to find further areas of interest in justice research, particularly as it relates to social, political and environmental studies. For, these chapters collectively offer other researchers a rich source of concepts, principles and theories, together with a sense of the ontologies that undergird them, the genealogies out of which they have arisen, and the ways in which they remain limited or partial discourses in need of further development. These genealogies are important because they show that while the whole point of justice theorising is to develop appropriately general, abstract and universalisable concepts, their origins and applications are always historically, geographically and socially situated. As researchers, students and policy makers work with the concepts and theories outlined in this volume, the historical, geographical and social situations in which they are operating will necessarily lead to the development and evolution of justice theory.

We hope that this diverse set of chapters has been understood with this purpose in mind.

Note

[1] Though not a complete or total coverage: for something closer to that, serious consideration also needs to be paid to matters of, for instance, mobility and food justice.

Reference

Okin, S.M. (1989) *Justice, Gender, and the Family*. New York: Basic Books.

Index